A Review on Recent Echocardiographic Software

Mersedeh Karvandi · Saeed Ranjbar

A Review on Recent Echocardiographic Software

Advancing the Field through the Emerging Science

Springer

Mersedeh Karvandi
Taleghani Hospital
Shahid Beheshti University of Medical
Sciences
Tehran, Iran

Saeed Ranjbar
Institute of Cardiovascular Research
Shahid Beheshti University of Medical
Sciences
Tehran, Iran

ISBN 978-3-031-29048-0   ISBN 978-3-031-29046-6   (eBook)
https://doi.org/10.1007/978-3-031-29046-6

This Springer imprint is published by the registered company Springer Nature Switzerland AG
The registered company address is: Gewerbestrasse 11, 6330 Cham, Switzerland

Preface

Recent technological developments have introduced new diagnostic devices to doctors, who can use the devices to diagnose diseases faster and more precisely. In addition, doctors can suggest better treatments using the new devices. However, since these developments merge basic sciences such as mathematics and physics with bioscience, understanding the basic and functional principles of these new devices is not easy. When the users—who are mostly doctors—understand all the functional aspects of the new devices, they can gain the best results from them, find new functions for them, and even promote them.

Echocardiography devices have a range of modern and sophisticated hardware components that have been manufactured based on mathematics, physics, and cardiovascular physiology; therefore, to get optimal usage from these devices and maximize their capabilities for diagnosis, all doctors should have a basic knowledge of how they function. That is why we decided to compile this manual, which introduces the details of sophisticated systems to doctors in a simple manner.

This manual is also useful for researchers who want to enter interdisciplinary fields such as physics, mathematics, medicines, and computer science. Finally, users of echocardiography devices such as applicable fellowships will benefit from studying this manual.

Tehran, Iran
Mersedeh Karvandi
Saeed Ranjbar

Contents

Echocardiography of Left Ventricular Myocardium Deformation

1

As we know, evaluating the deformation of the myocardial muscle in the cardiac patient is important. Evaluating the myocardial motion includes evaluating the velocity and its displacement size, and evaluating the deformation includes evaluating the strain and strain rate.

The strain is the same as the muscular deformation compared to its primary shape, and the strain rate is the velocity of this muscular deformation. The measured type of deformation by echocardiography machines is based on Lagrangian deformation; however, the various systems of these machines can identify the deformation indexes (strain and strain rate) differently.

Assigning the deformation at the various segments of the left ventricular myocardial muscle should be performed by experts, but assigning systems in echocardiography devices have been designed that can automatically measure the deformation indexes.

The "strain" parameter can be measured via different mathematical techniques, such as tissue Doppler imaging (TDI) combined with velocity gradient (VG), speckle tracking (ST), speckle tracking with seven endpoints, speckle tracking combined with TDI, 2D base strain (which means speckle tracking combined with velocity vector imaging [VVI]), and left ventricular (LV) modeling, as a polygon shape combined with speckle tracking and even simple M-mode is able to measure a special kind of strain. Therefore, the value of strain depends on how one measures the strain and what kind of strain is meant (longitudinal, radial, circumferential strain, etc.). The following review provides a general overview of the currently available imaging techniques that enable strain measurements based on the available literature and ongoing works [1–3].

The main goals in this chapter are to review the measurement methods of the velocity and deformation indexes (left ventricular myocardium) in the different echocardiography machines and to cover the methods of their mathematical software.

M. Karvandi, S. Ranjbar, *A Review on Recent Echocardiographic Software*,
https://doi.org/10.1007/978-3-031-29046-6_1

These methods are as follows:

TDI can indicate the velocity in any segment of the myocardium and calculate the displacement size and the applicable segment's deformation size (strain and strain rate). The main question is, how can the velocity at the applicable cardiac muscular tissue be calculated by the tissue Doppler ultrasound method?

If we calculate any segment at the left ventricle, such as the septal base, the reversal waves from this segment in the time, T_0, would be a frequency of f_0 (Fig. 1.1).

Despite being dynamic, all reflected waves in time T_1 are a wave with the frequency f_1. The applicable segment in a cardiac cycle from T_0 to T_n could have different frequencies from f_0 to f_n (Figs. 1.2 and 1.3).

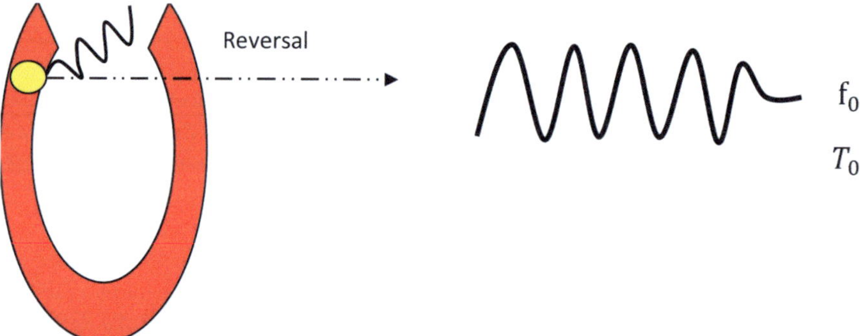

Fig. 1.1   Reversal wave of cardiac segment

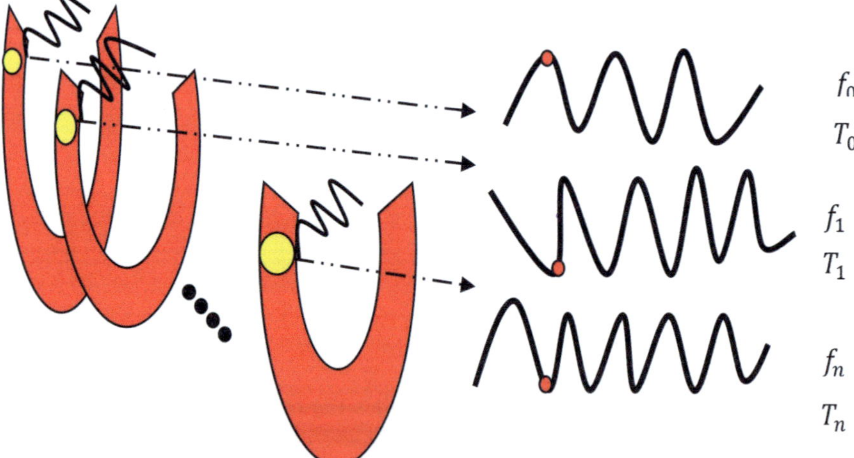

Fig. 1.2   Reversal waves from the basal septal segment in a cardiac cycle; all reversal waves are disrupted by each other at different points

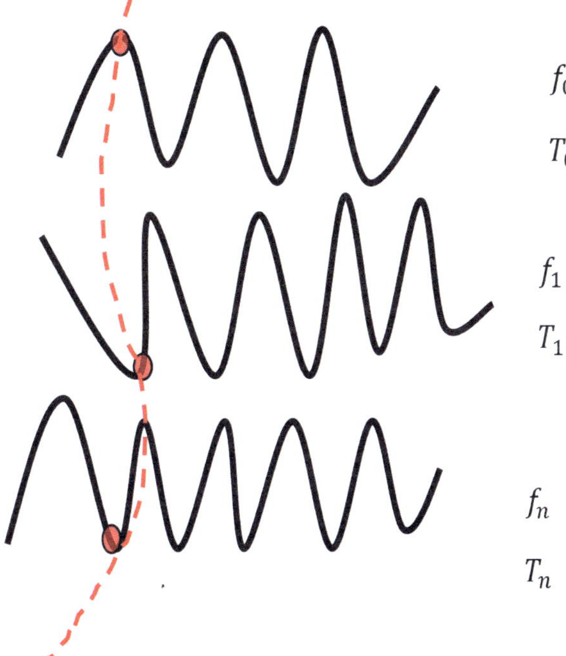

Fig. 1.3 Waves' disruption at different points

If we connect these points to each other, we will have a curve that indicates the waves' disruption; this curve is called the *wave equation of d'Alembert* (Fig. 1.4).

This equation can be used to get the velocity curve via a Fourier transform [4, 5] (Fig. 1.5a, b).

From the velocity curve of each segment in the left ventricle, we can assign the strain and strain rate. For instance, if the septal base in T_1 has velocity v_1 and in T_2 has v_2, and from T_1 to T_2 has a displacement of size L, then the strain rate would be calculated as follows:

$$\varepsilon'\left(L_1,t\right) = \frac{V_2\left(t\right)-V_1}{L\left(t\right)}$$

From the integral of the strain rate, we could get the strain as follows (Figs. 1.6 and 1.7):

$$\varepsilon\left(L_1,t\right) = \int_0^t \varepsilon'\left(L_1,u\right)\mathrm{d}u$$

Fig. 1.4 Wave curve of d'Alembert

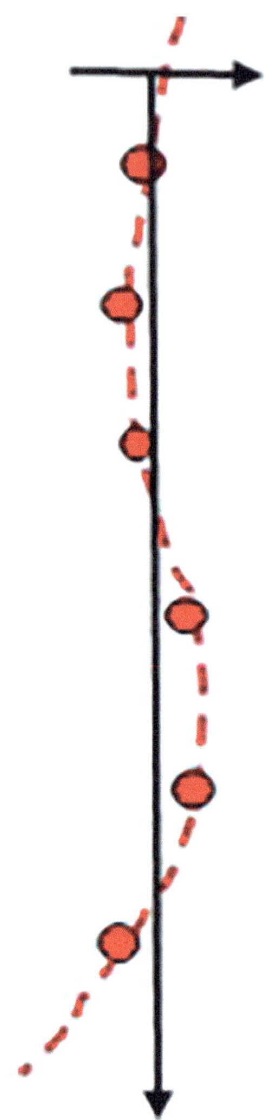

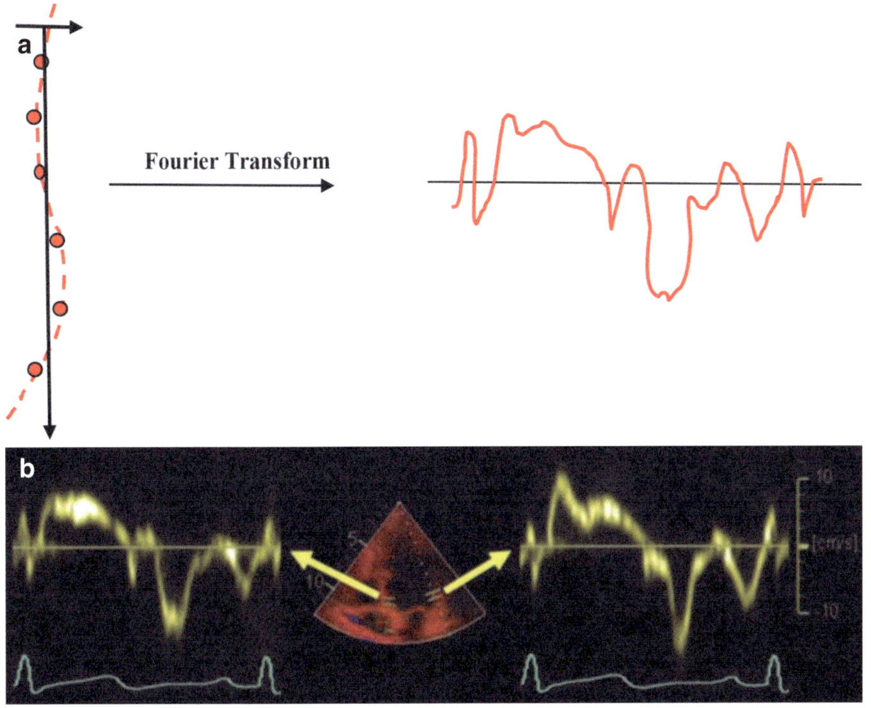

Fig. 1.5 (**a**) Velocity curve in cardiac muscle, (**b**) velocity of myocardium based on the TDI method

Fig. 1.6 The TDI-VG method to obtain the strain by the time integrating the strain rate curve. V_1 is the velocity at the region of interest 1, and V_2 is the velocity at search region 2. L is the distance between the 1 and 2 regions

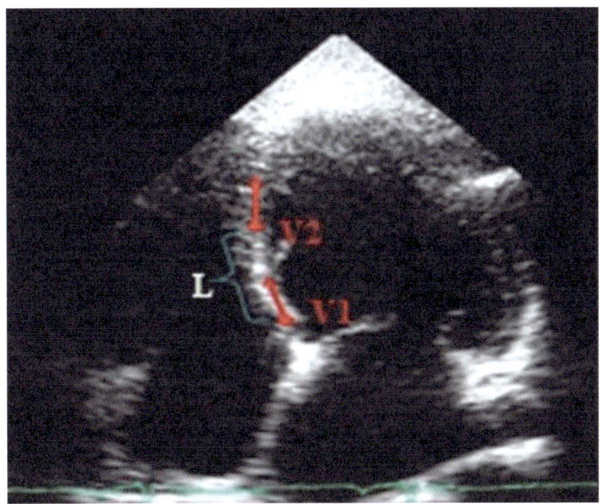

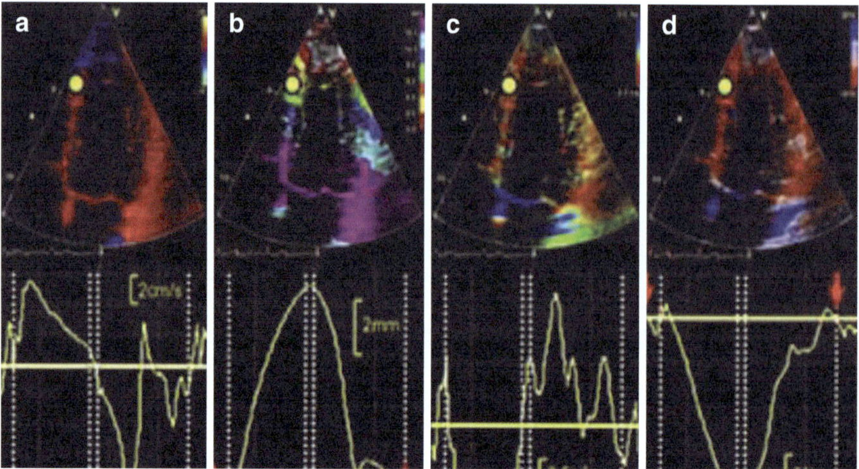

Fig. 1.7 (a–d) Velocity and displacement of myocardium and deformation indexes (strain and strain rate) based on TDI method

Speckle Tracking Method

In the echocardiography machines with the speckle tracking system, the velocity and tissue moving would not be assigned by the tissue Doppler ultrasound imaging method. In this system, there is a new software in the field of image processing that can assign the velocity and deformation of the cardiac muscle. This software can track the applicable segment and can identify its place on the coordinate screen.

In this system, the strain would be calculated first, and then the strain rate would be calculated. For instance, the basal septal segment would be tracked in time T_0 at point A with the clear coordinate (X, Y, Z), in time T_1 at point B, and in time T_n at point N (Fig. 1.8).

Points A, B, ... , N would be connected by vectors (Fig. 1.9).

The aggregated number of these vectors' length as the applicable segment reaches from A to N could calculate the strain as well.

$$\varepsilon\left(i,t_n\right) = \sum_{k=1}^{n-1} \frac{\left(L_{k+1,t_{k+1}}, - L_{k,t_k}\right)}{L_{k,t_k}}$$

From the differentiation of the strain, we could calculate the strain rate, and from the differentiation of displacement, we could calculate the velocity (Fig. 1.10a–c).

So, in the TDI method, the velocity, displacement, strain rate, and strain would be respectively calculated, and this trend in the speckle tracking would be performed as follows: displacement, strain, strain rate, and velocity.

Fig. 1.8 Points' status on coordinate screen

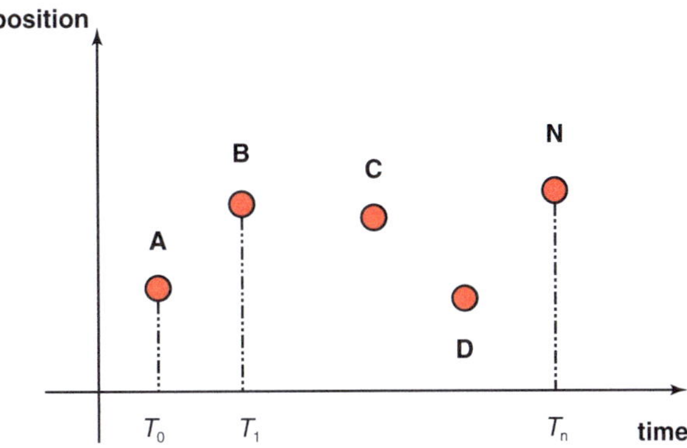

Fig. 1.9 Connecting vectors of points and vector of total displacement (red vector)

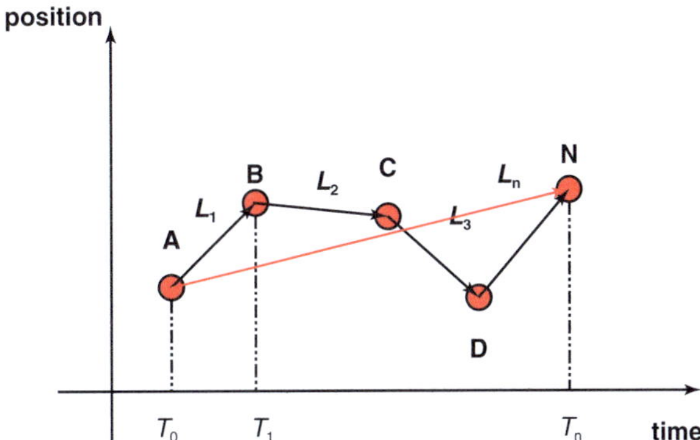

Merging the Speckle Tracking and Tissue Doppler Imaging

In this method, the moving location of the applicable segment in a cardiac cycle will be tracked first by the ST software (Fig. 1.11).

By contrast, with speckle tracking, the length of line segment between the A, B, C, ... , N points from T_0 to T_1 could have been calculated to assign the strain; however, in this method, the change of displacement from A to B is done with the TDI method.

Hence, we can get the velocity of the applicable segment in A and B from T_0 to T_1. Next, the hatching location under curve is assigned. This location is the same as the displacement change, and its symbol is L_{AB} (Fig. 1.12).

For instance, if we want to have the strain of the basal septal segment in the interested time (t), then the strain would be calculated as follows.

If this segment in T_0 is located at point A, and t is located at point K, then it has passed through B, C, D, ... , J, so that it will reach the same point as K (in applicable time of t). Therefore, we can say the following:

$$\varepsilon\left(A,T_i\right) = \sum_{n=1}^{i} \frac{\left(L_{AN,n}\left(T_i\right) - L_{A(N-1),n-1}\left(T_{i-1}\right)\right)}{L_{A(N-1),n-1}\left(T_{i-1}\right)}$$

AK in the formula, the displacement length, is from A to K.

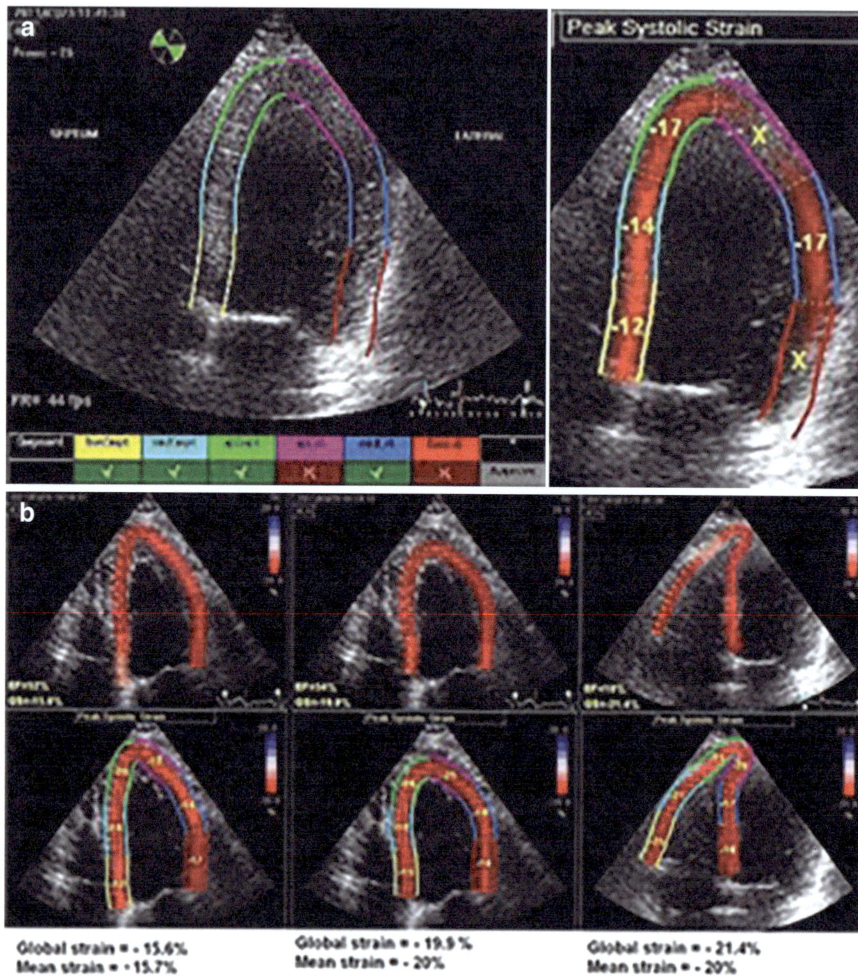

Fig. 1.10 (**a**) Strain of myocardium based on speckle tracking. (**b**) Longitudinal strain by speckle tracking. (**c**) Radial strain by speckle tracking

Fig. 1.10 (continued)

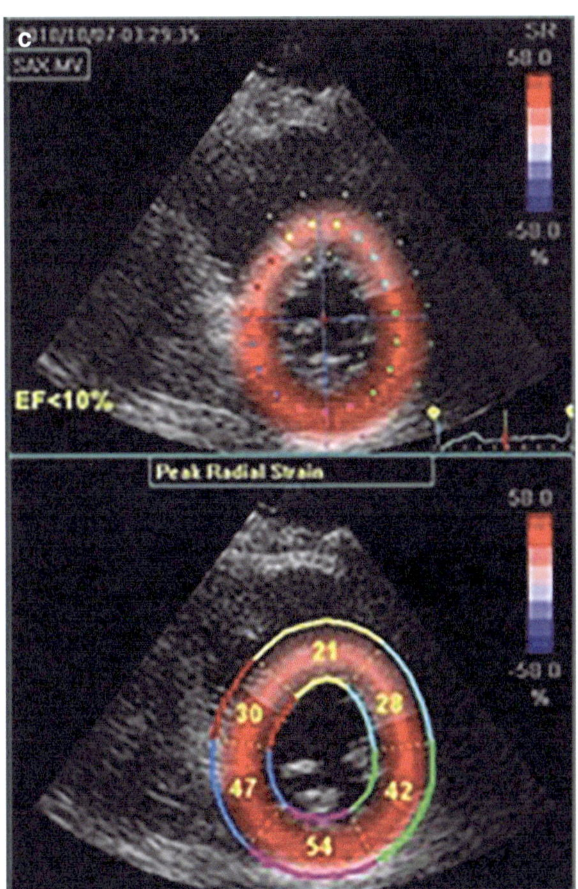

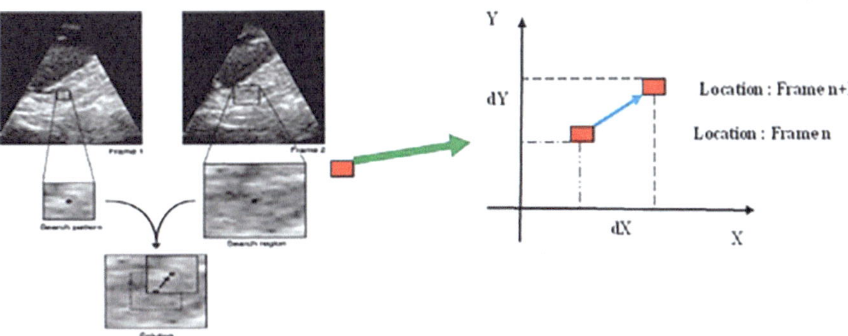

Fig. 1.11 Assigning the location of segment with the 2D ST software

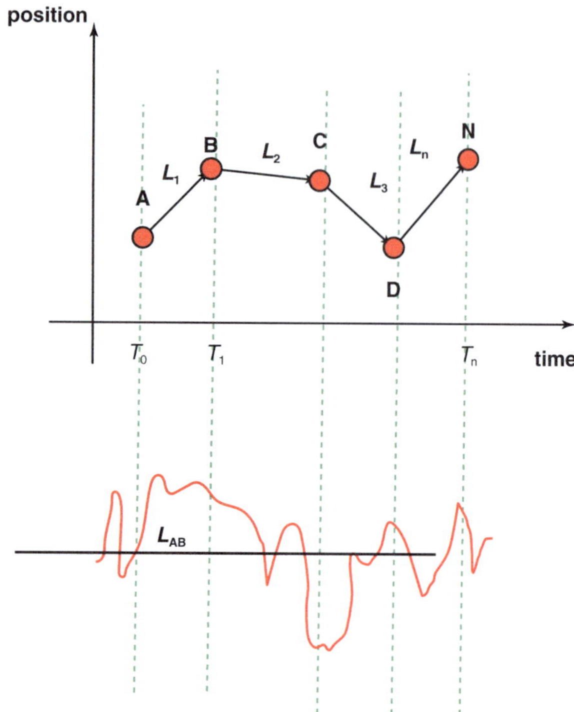

Fig. 1.12 Merging 2D ST and TDI methods. The red curve is the velocity of applicable segment by the TDI method, and L_{AB} is the enclosed location in the statement from T_0 to T_1

Velocity Vector Imaging Algorithm (VVI)

- Echocardiographic images are acquired from LV 4 apical chamber (4C) views.
- $A_{1,k}$ is the region of interest of the left ventricular myocardium manually divided into 13 points at 4C (1th frame): $k = 1, 2, 3, \ldots , 13$ (Fig. 1.13).
- Figure 1.14 shows perpendicular lines crossing each divided point ($k = 1, 2, 3, \ldots, 13$).
- $A_{2,k}$: New points automatically identified from the intersection between perpendicular lines and the second frame ($k = 1, 2, 3\ldots, 13$).
- $A_{i,k}$: New regions using the same procedures as the previous steps for the ith frame (from 1th frame to ith frame).
- T_i: Time corresponding to ith frame.
- $l_{i+1,T_i,k}$: Distances between $A_{i+1,k}$ and $A_{i,k}$ corresponding to T_i for each $k = 1, 2, 3, \ldots , 13$.
- $\varepsilon\left(A_{1,k},T_{i+1}\right) = \sum_{j=1}^{j=i+1} l_{j+1,T_{j+1},k} - l_{j,T_j,k} / l_{j,T_j,k}$: Strain value of points $A_{1,k}$ for each k at time T_{i+1}.

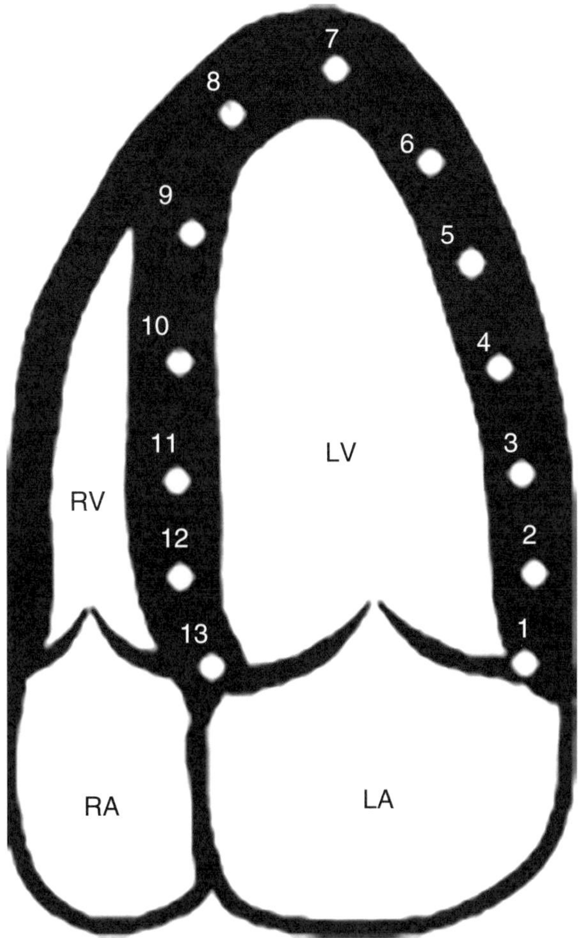

Fig. 1.13 The LV myocardial region has been divided into 13 points in a 4C view

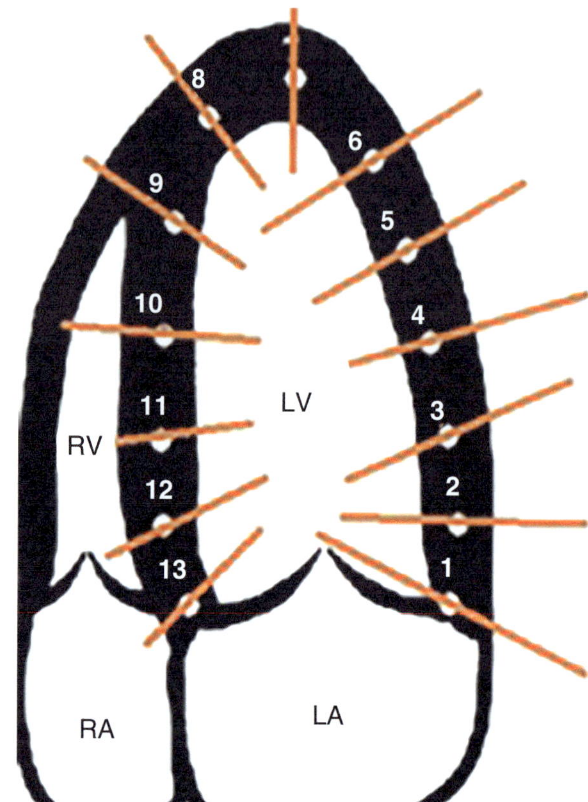

Fig. 1.14 Perpendicular lines crossing all 13 points

2D Base Strain Method

In this method, like the previous methods, the moving location of the applicable segment in a cardiac cycle is tracked by the ST software.

The main question here is, which part of this segment is tracked as a sample? In the previous methods, the core part was considered as the sample, its displacement assigned the strain, and the rest of the indexes of deformation were considered as well (Fig. 1.15).

In this method, the applicable segment is considered as a complex of the points, and any segment includes a large number of points, but these points have a special order in their dispersion and distribution against the core point.

In fact, the distribution average of these points is the core point of the applicable segment. In other words, the central kernel is the distributive average of its own peripheral points (Fig. 1.16a, b).

In fact, 2D base strain software is based on the scattering distribution; therefore, the ST software first determines the applicable segment (Fig. 1.17).

Next, the SP software determines the central kernel and its coordinate position (Fig. 1.18).

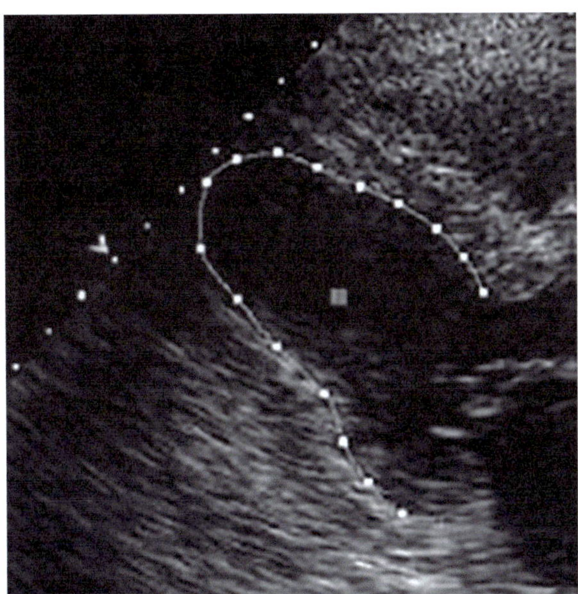

Fig. 1.15 Considered core segments in previous methods

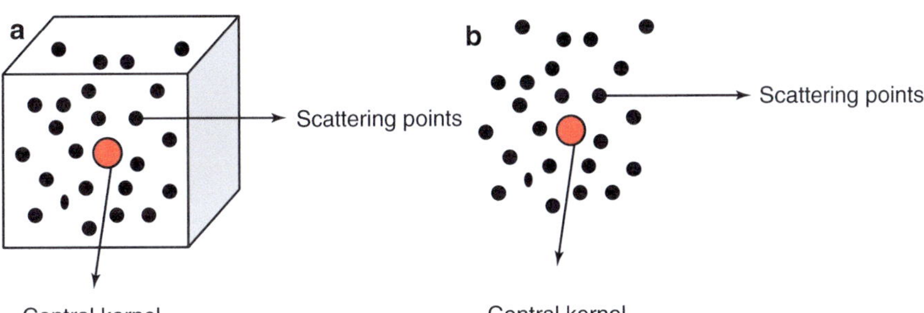

Fig. 1.16 (**a**) Central kernel as an average from its peripheral scattering points. (**b**) Scattering points around the central kernel on the coordinate screen

The central kernel (μ) has a and b sizes as the distributive average of the inner points. It is worthy to note that, in the applicable segment, the points that are scattering around the central kernel will be utilized.

In a cardiac cycle, the central point and its peripheral points are displaced, and in each displacement, based on time unit, the arrangement of these peripheral points is again the distributive average of these points in the name of the central kernel (Fig. 1.19).

The main question is, how is the placement of these points in the segment determined? The computer software can use a volume unit as a circle in which there are smaller circles with 1.2, 1.4, … , $1/n$ as the radius, and the center of all these circles will be the same central kernel of applicable segment (Fig. 1.20).

Despite that the radius of the smallest circle is $1/n$ alongside $1/n \to 0$, the smallest circle is the same central kernel. Meanwhile, the software determines the center

Fig. 1.17 A segment with determined size

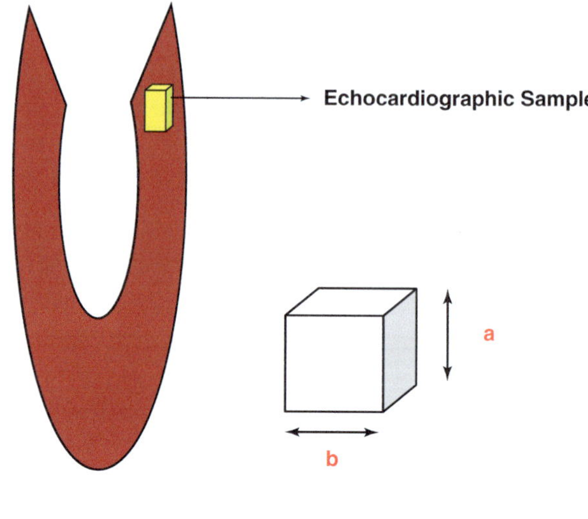

Fig. 1.18 Size and central kernel of segment

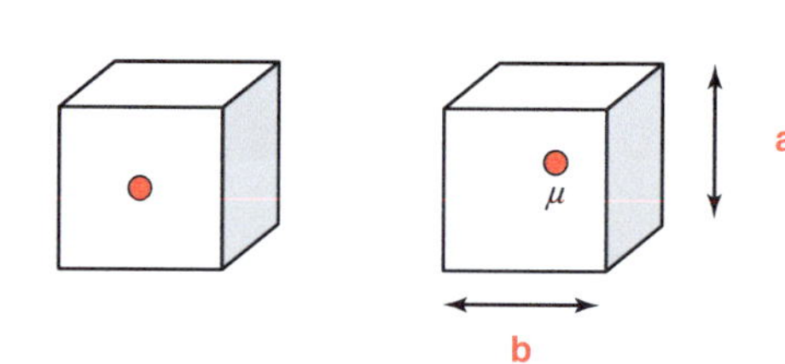

Fig. 1.19 Displacement of a central kernel and its peripheral scattering points from a frame to another

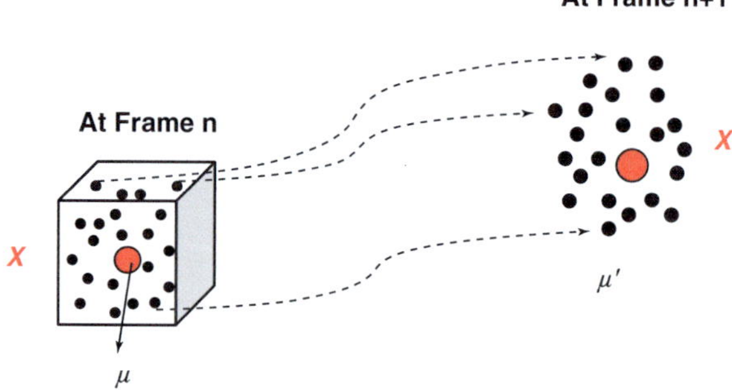

Fig. 1.20 Applicable points around the central kernel of a segment

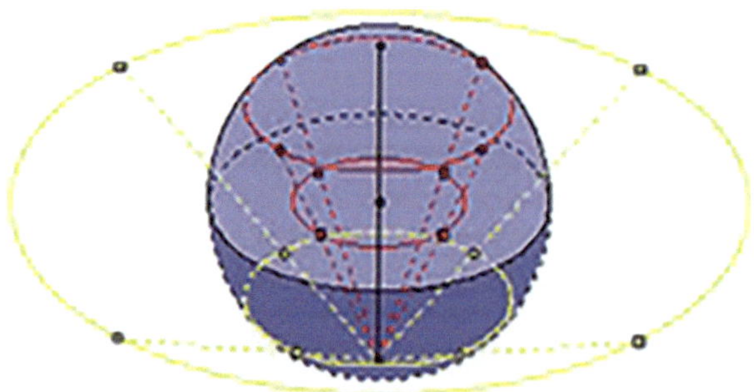

and coordinate location of the applicable points. Hence, we can find the density function of the points' distribution with the following:

$$F(t) = 1/\sqrt{2\pi}\sigma \int_0^t e^{(X-\mu/\sigma)x}\,dx$$

In fact, X is known as the applicable points, μ is the distribution average of points, and σ is the variance of scattering points. In frame $n+1$, by including nverrse $F(t)\,F^{-1}$, we get the new points with a center of μ. In fact, μ' can be tracked by the ST software as with the past methods.

The 2D base strain system shows that based on the $X' = \log_e(F^{-1})$ function, F is the density function of the applicable distribution, and $\log_e$ is a natural logarithm based on the number of e. Therefore, the new location could determine the applicable points in the next frame $(n+1)$ and in time T_{n+1} (Fig. 1.21).

After determining the points' locations of a segment in the frame, we can assign the displacement distance to the same points. Finally, the strain of applicable segment will be determined in the cardiac cycle.

The main question is, how is the distance length determined? In fact, not only is this distance length a straight line, but the two points can be connected in different ways. Each function such as f has a graph or shape. For example, the applicable point A, which was determined by the assumed function f to have reached point B, in fact could move to the graph of the assumed function f (Fig. 1.22). So, the distance between $A - B$ would be calculated as follows:

$$L_{AB} = \int_a^b \sqrt{f'}\,dx$$

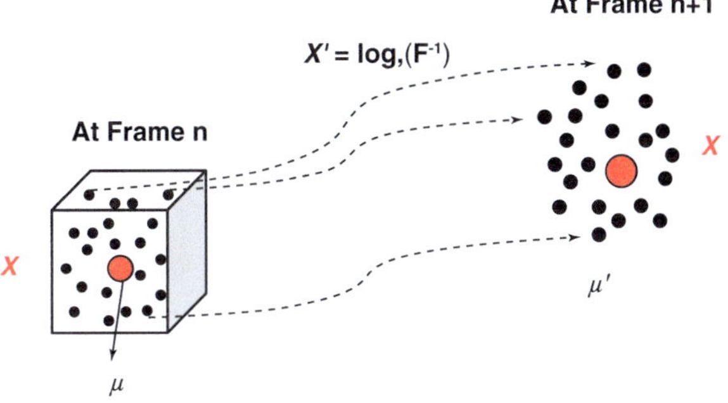

Fig. 1.21 Displacement of a central kernel and its peripheral scattering points from one frame to another using the $\log_e(F^{-1})$ function

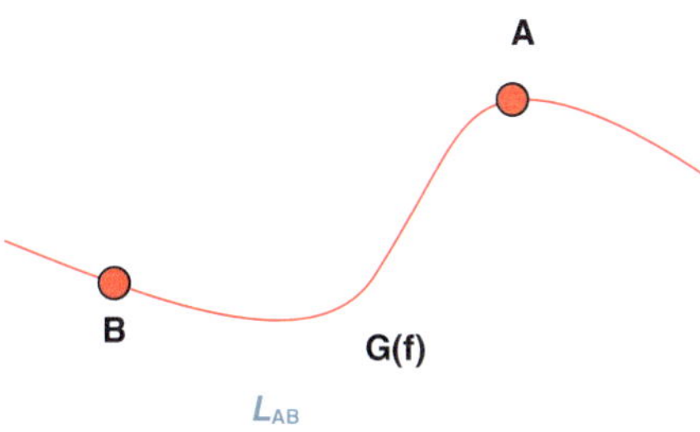

Fig. 1.22 Passing function curve f through the points $A - B$

The graph of the applicable function $X' = \log_e(F^{-1})$ is as shown in Fig. 1.23.

If we take the center point of the applicable segment and its own assumed peripheral points a, b, and c, then in $T + 1$ the new location of the central kernel u' based on the SP software and a', b', and, c' points based on the logarithmic function are assigned.

The distance from a to a' , b to b',and c to c' will be assigned using the same formula.

For example, if we want to determine the strain in a basal septal segment, then first we should determine the location of the central kernel and its peripheral points at T_1. Then we determine the points' locations at the next frame and determine T_2 by using $X' = \log_e\left(F_{\mu_1}^{-1}\right)$. Finally, we determine the location of the central kernel and its peripheral points at frame n and determine T_n using $Ln\left(F_{\mu_{n-1}}^{-1}\right)$ [6–8] (Fig. 1.24).

We can determine the distances average of the strain by using the 2D base vector velocity imaging (VVI) method (Figs. 1.25 and 1.26).

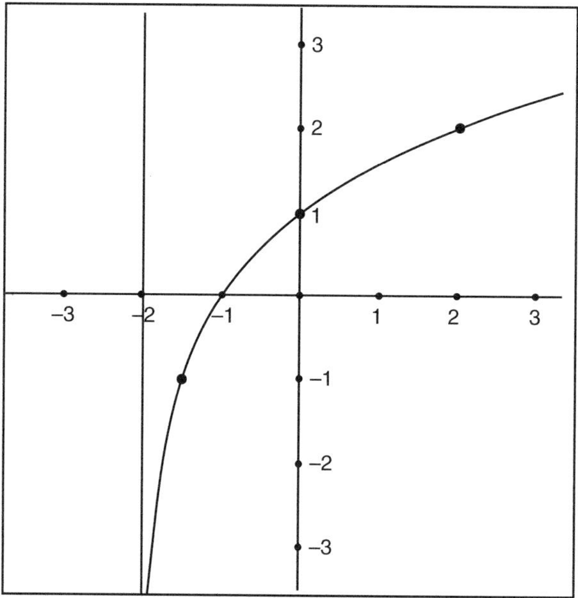

Fig. 1.23 Graph of logarithmic function $X = \log_e(F^{-1})$ on the coordinate screen

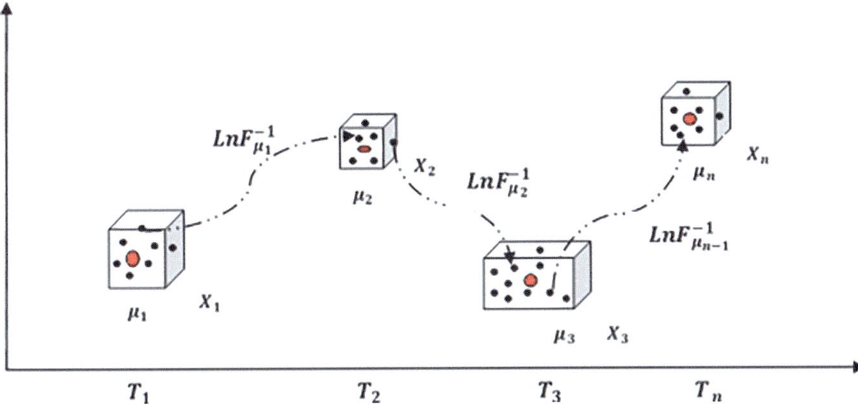

Fig. 1.24 Tracking a myocardial segment using the X-Strain method or vector velocity imaging

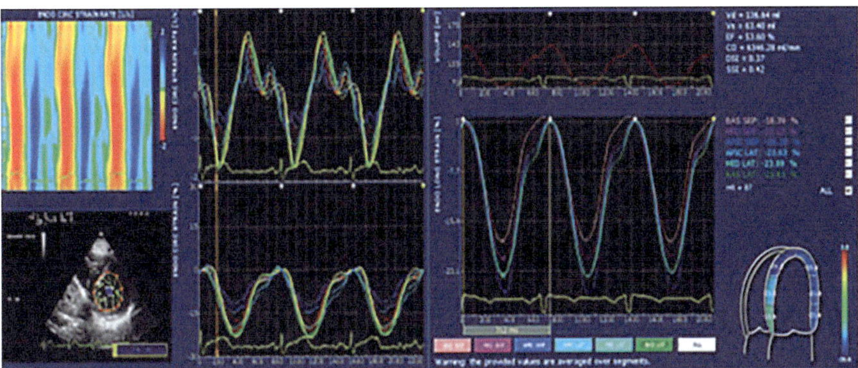

Fig. 1.25 Numerical and diagram consideration of longitudinal strain based on the 2D base strain (right) and circumferential strain using the 2D base strain (left)

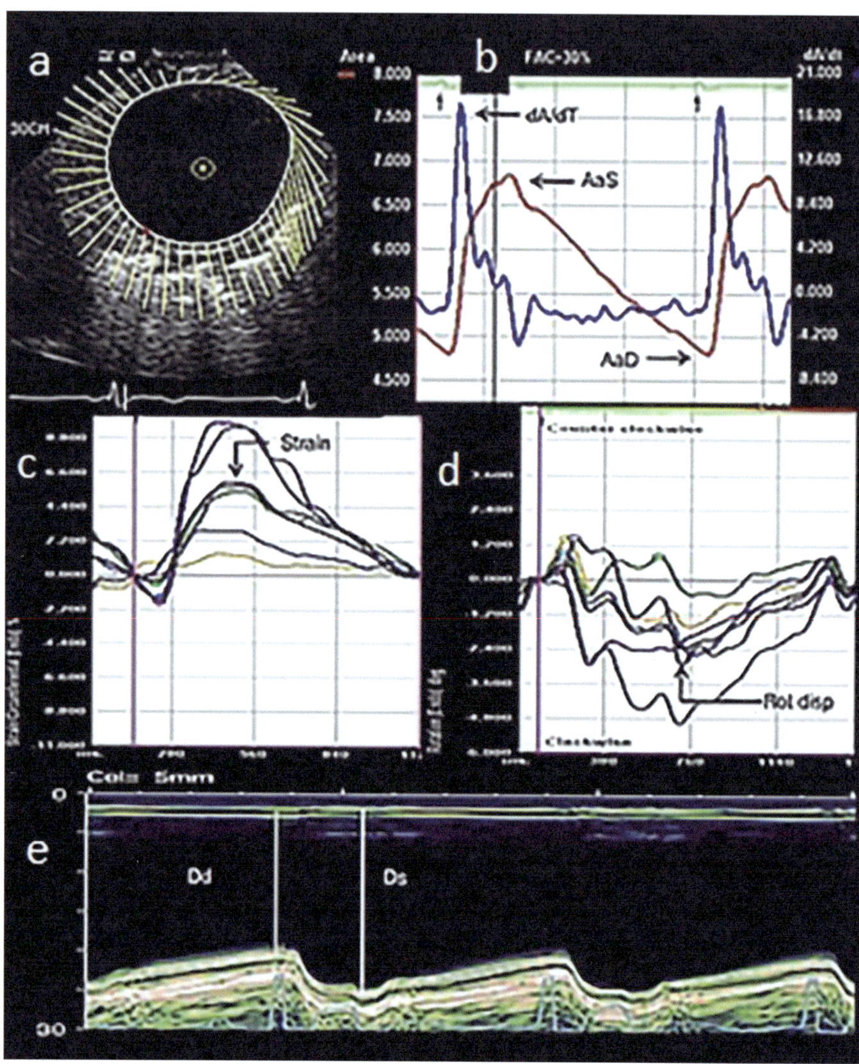

Fig. 1.26 Considering the radial strain based on the 2D base VVI method (**a–e**)

3D Echocardiography Systems

In echocardiography systems and machines, to consider the deformation of the left ventricular muscle, we can use the 3D speckle tracking software. In these machines, the left ventricular segment is used as a volume element (Fig. 1.27).

In the 2D ST software, based on the introduced model for LV, the velocity and displacement of the applicable point (in cylindrical LV) or network (in polygonal model LV) can be considered. In the 3D ST software, a displacement in the equilateral shape is the main subject. Note that, in both models, calculating the velocity

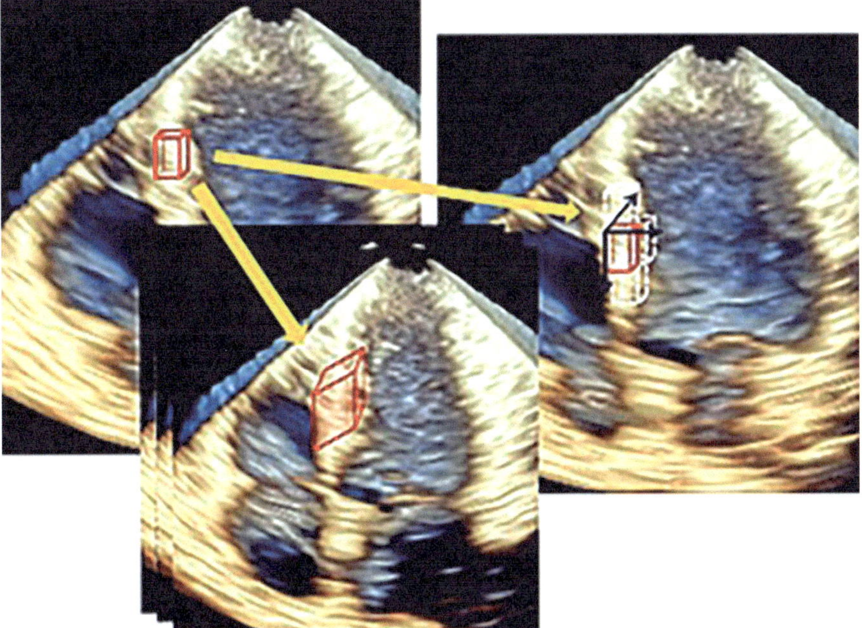

Fig. 1.27 Segment of left ventricular muscle as an equilateral shape in 3D ST software

Fig. 1.28 Diameter and width in equilateral shape

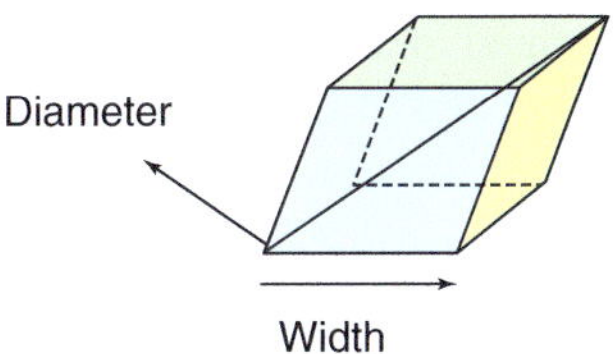

and deformation of the myocardial segments is performed on three axes: longitudinal, radial, and circumferential. In fact, in the 3D ST software, this deformation in a cardiac cycle is calculated as the change percent of the diameter in the equilateral shape against its width (Fig. 1.28).

The 3D strain of a segment in 3D ST is calculated as follows:

$$3D \ Segmental \ Strain = \frac{L - L_0}{L_0}$$

The previous diameter and width in the equilateral shape are L and L_0, respectively (Fig. 1.29).

Fig. 1.29 Diameter and width of a left ventricular segment

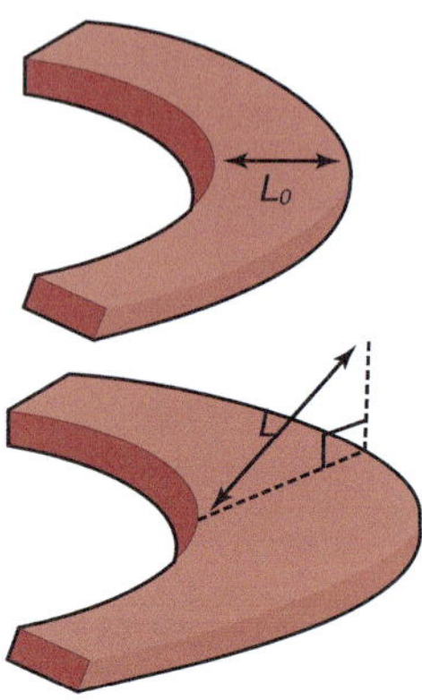

4D Echocardiography Systems

4D strain software is applied on the 2D systems using the left ventricular cylindrical model. To have the 2D echocardiography system use the 3D images, first the images are acquired in 4C and 2C apical views and in long-axis, short-axis views. Then the region of interests are determined by the ST software. In fact, a range of the mesh screens can be achieved alongside these images in which each network includes a left ventricular segment, and when these networks (mesh) are connected, a 3D model of the left ventricle will be created. Since this modeling trend can be performed in a cardiac cycle, a 4D dynamic model of the left ventricle will emerge as well (Figs. 1.30, 1.31, 1.32 and 1.33).

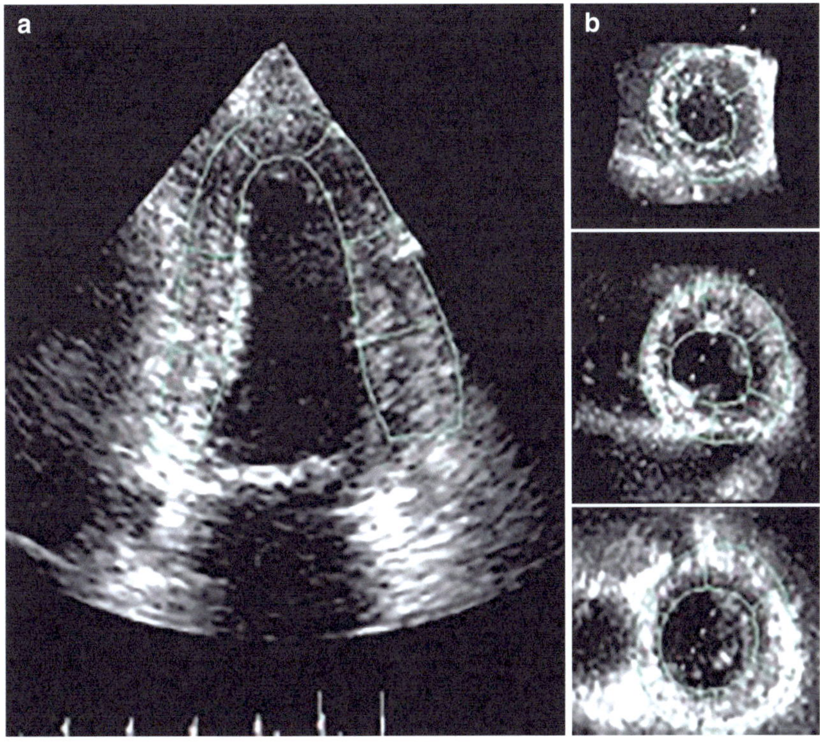

Fig. 1.30 Segmentation of different views of the left ventricular muscle in short-axis and 4C by ST software (**a**) for long axis view and (**b**) for short axis view

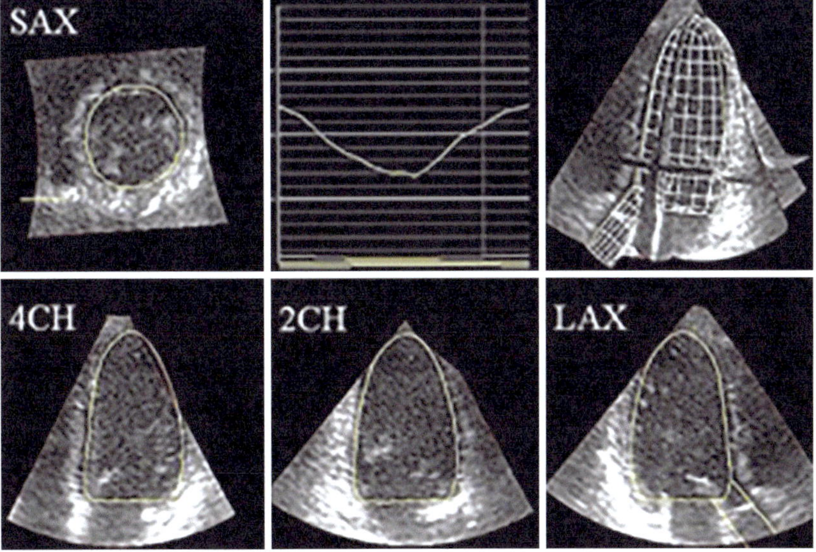

Fig. 1.31 LV in different views

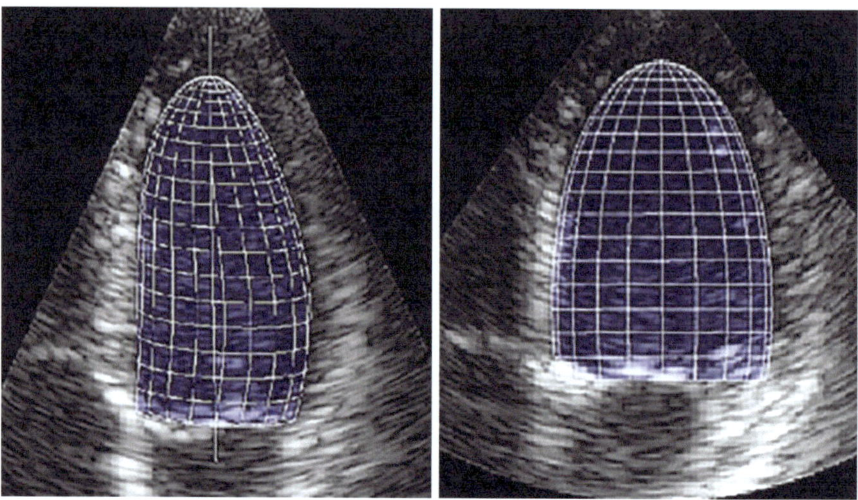

Fig. 1.32 3D echocardiography picture using 4D technology

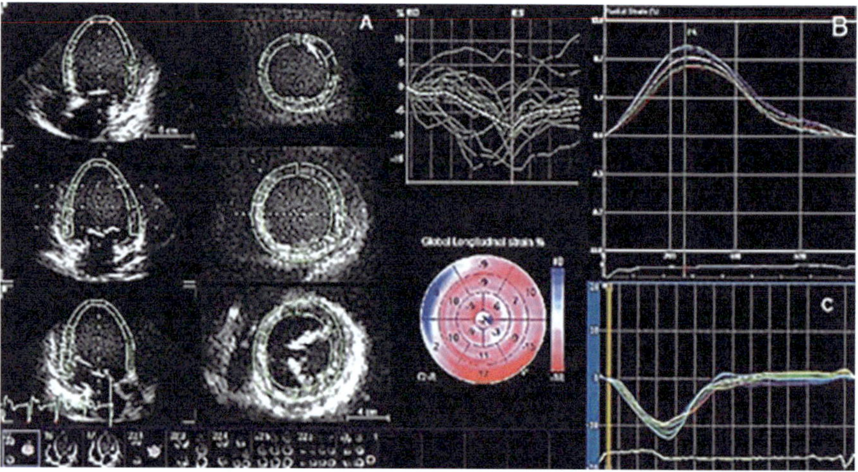

Fig. 1.33 Strain diagrams in direction of longitudinal, radial and circumferential (A, B and C respectively) at 4D system

Reconstructing the Right and Left Ventricular Layers with the TomTec Software

This software can reconstruct the cardiac cavities using a new method. This reconstruction will principally happen by triangulating the right and left ventricular cavities.

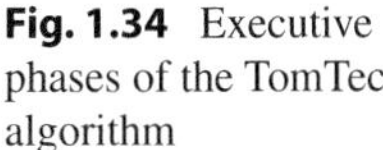

Fig. 1.34 Executive phases of the TomTec algorithm

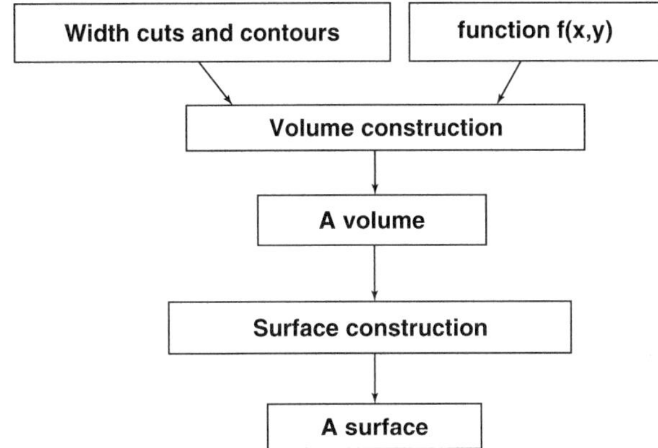

This trend can provide better and more exact information from the ventricular indices such as deformation, etc. In this software, the sectional cuttings can border the 3D images of both ventricles. These borders can be connected by some of the functions, and finally a layer model of 3D image can be achieved (Fig. 1.34).

As mentioned, the main goal of this method are to create a 3D layer using contours. This layer is a smooth one; all sharp and jagged points are not considered.

The volume construction can be achieved by using the S_1, S_2, S_{N_z} contours and the $f(x, y)$ function. This function could be presented on all screens having the following contours:

$$f(x,y) = \begin{cases} < 0 \; if \; (x,y) \; is \; outside \; all \; contours \\ \quad 0 \; if \; (x,y) \; is \; on \; a \; contour \\ > 0 \; if \; (x,y) \; is \; in \; side \; a \; contour \end{cases}$$

The mesh-like network $N_x \times N_Y$ on the $y - x$ axis can be uniformly placed on the contours (Fig. 1.35).

By including the z-axis in the vertical direction on the meshed contours, a 3D network is built in which all points are cubical pixels (Fig. 1.36).

The algorithm of the TomTec software follows a range of the points on the 3D network so that $f(x, y, z)$ is zero. So, intersecting the points or the same cubical pixels could be triangulated as the applicable layer (Fig. 1.37).

Now, the question is, how is $f(x, y, z)$ exercised? First, a simple function with the same name as the decision function will be used. This function can attribute -1 to the outer point of the layer (surface), 1 to the inner point, and 0 to the points that at least are on a contour (Fig. 1.38).

For the two adjacent contours S_{i+1}, S_i from a layer (surface), the related numbers are -1, 1, and 0, and also the three different networking will be drawn (Fig. 1.39a, b).

The distance function to layering is much better than the decision function, as shown here:

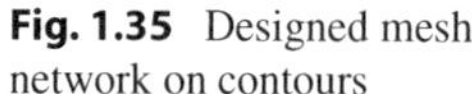

Fig. 1.35 Designed mesh network on contours

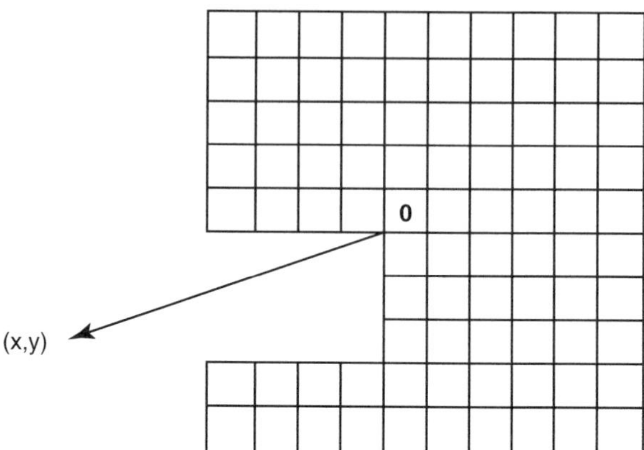

Fig. 1.36 Networking of 3D one-layer contours

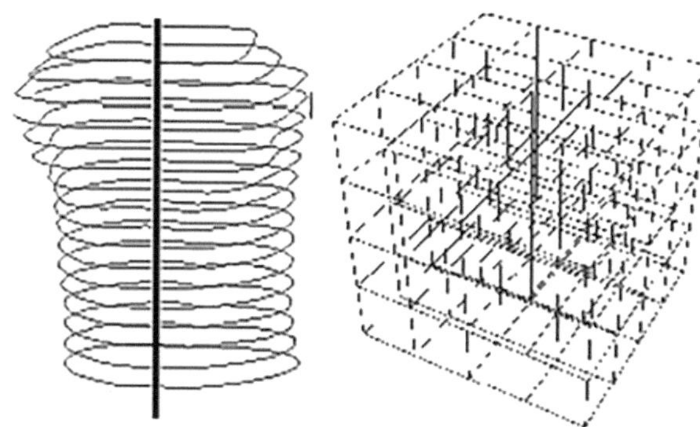

$$f(x,y) = \begin{cases} -dist(x,y) & if\ (x,y)\ is\ outside\ all\ contours \\ 0 & if\ (x,y)\ is\ on\ a\ contour \\ dist(x,y) & if\ (x,y)\ is\ inside\ a\ contour \end{cases}$$

In fact, dist (x, y) is the closest distance of contour to (x, y). Then, the consistency of this function in the changes compared to the decision function will lead to a better approximation of the applicable layer (surface) (Fig. 1.36a, b).

With the two decision and distance functions' performance on all the networked contours, the applicable layer will be created (Fig. 1.40a, b).

As noted earlier, the 3D networking on some of the contours will be made through the cross-sectional cuttings. The points in this network are like the cubical pixels with the 26 intersected points as well (Fig. 1.41).

The two parameters will be accrued to each point: the point's position against the layer, and the point's distance to the layer. If this point is outer, inner, or on a border, it will be determined, respectively, as follows: -1, 1, and 0. For a pixel in the 3D network, the applicable function is as follows:

$$f(x,y,z) = \sum_{k=z-1}^{z+1} \sum_{j=y-1}^{y-1} \sum_{i=x-1}^{x+1} state\ field\ of\ voxel\,(i,j,k)$$

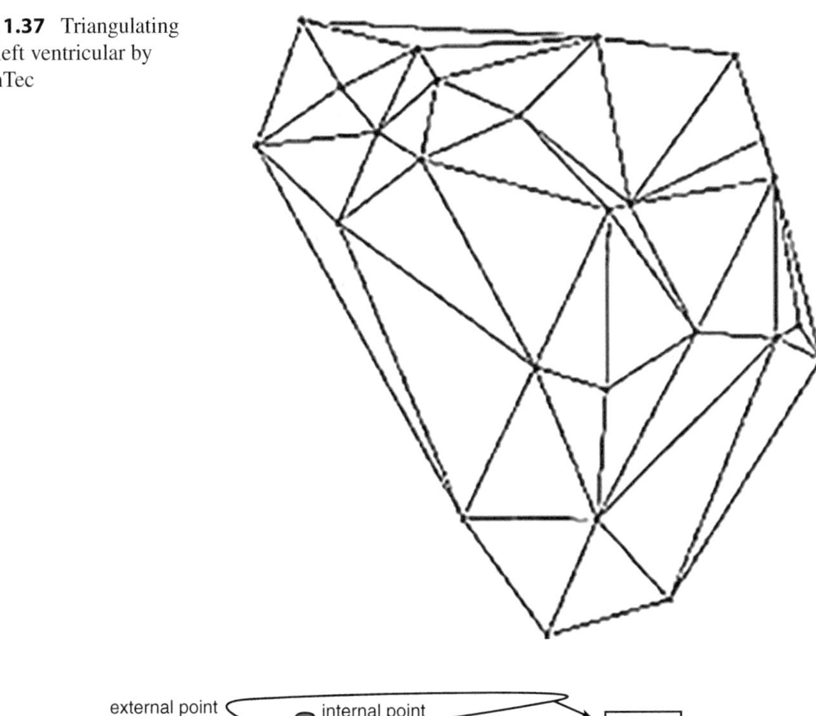

Fig. 1.37 Triangulating the left ventricular by TomTec

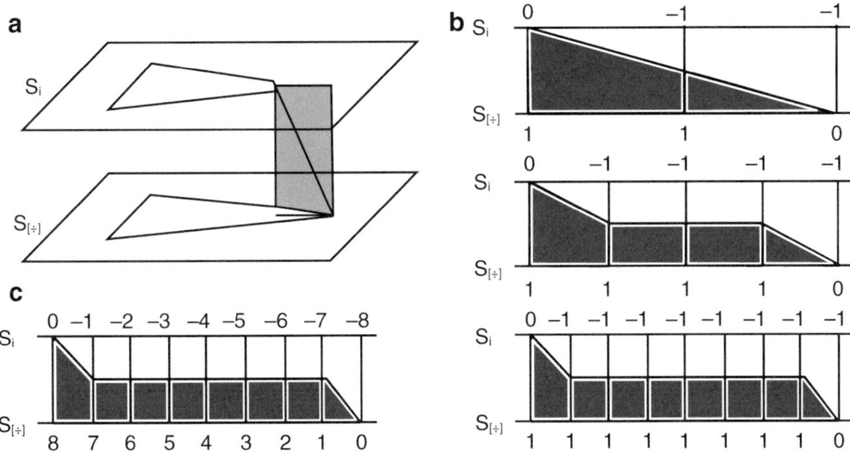

Fig. 1.38 Performance of decision function on the internal, external, and border points

Fig. 1.39 (**a**) Two adjacent contours; (**b**) Graphs of the decision function despite the three different networks; (**c**) Graphs of the distance function

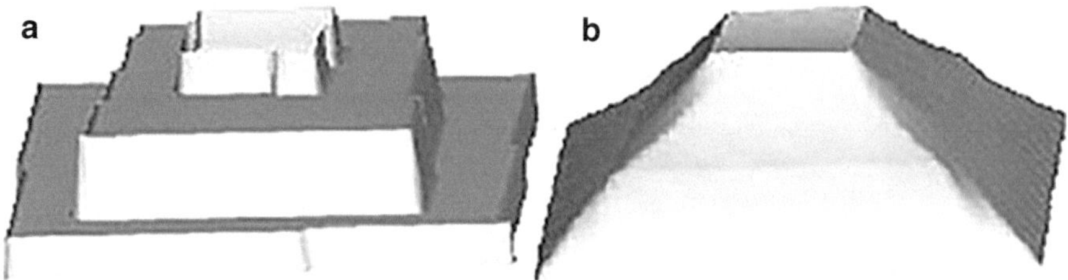

Fig. 1.40 (**a**) Reconstructing the layer using the decision function; (**b**) Reconstructing the layer by distance function

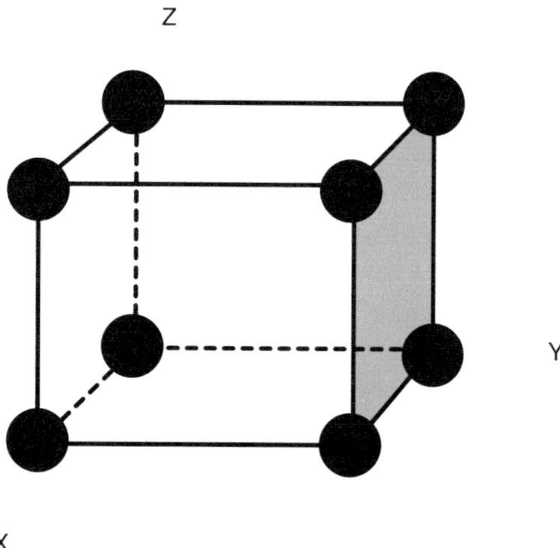

Fig. 1.41 x, y, z as consisting points of the 3D network have a maximum of 26 intersected points with the applicable layer (surface)

Based on this function, the point of a pixel on layer 0 will have an outer of -27 and an inner of 27.

The steps are as follows:

1. The cross-sectional cuttings establish the contours and mesh (networking) 3D ($N_X \times N_Y \times N_Z$) on the applicable layer.
2. Calculate the location function's values of network points.
3. Calculate the distance from the function's values, which are assigned a range of the points from this network (mesh) that have conformity with the applicable layer. This conformity and commonality with the applicable layer can create a new triangulated layer. Therefore, the right and left ventricles are reconstructed as a triangle and achieve the same contours from the cross-sectional cuttings.

First, in the short-axis view, the LV and RV contours are determined. The intersections of the right and left ventricular contours are assigned as r_1, r_2, r_3, r_4 points by the location function, and also the distance function determines the corresponding points of e_1, e_2, e_3, e_4 on the LV myocardium. The e_i points on the LV endocardium are the closest points to the r_i corresponding points (Fig. 1.42a), and the other points from the midpoints of the myocardial muscle are determined too (Fig. 1.42b).

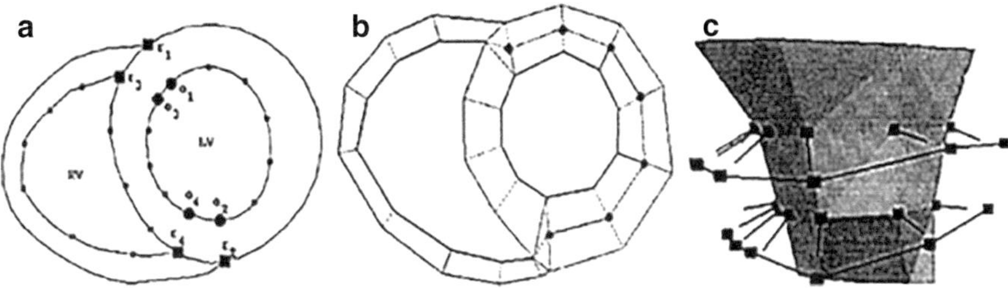

Fig. 1.42 (**a**) Zoning of a contour in short-axis; (**b**) Midpoints of LV myocardial muscle; (**c**) Connection between the sample points in the inner muscle with their corresponding ones in epicardium and triangulated picture from LV and RV

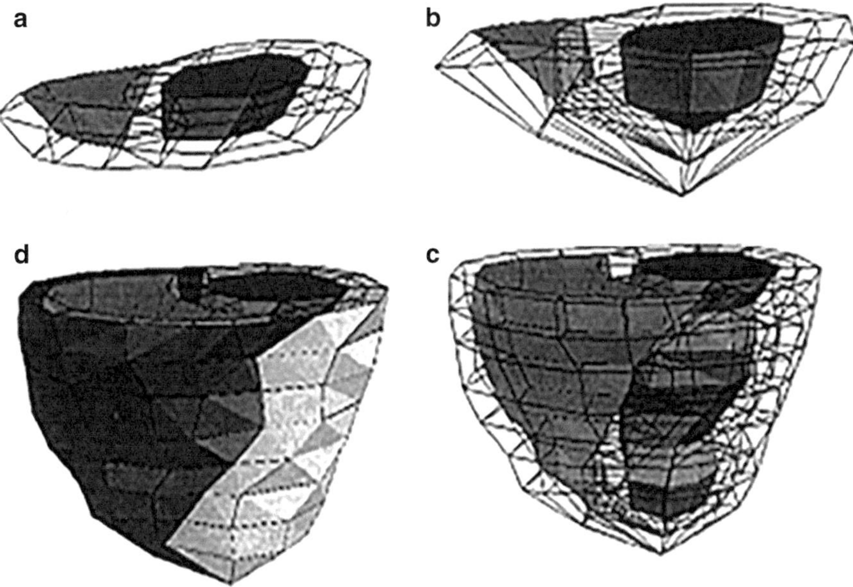

Fig. 1.43 Step-by-step model of triangulating the LV and RV layers (**a–d**)

The LV and RV contours based on the built 3D points can construct the right and left ventricular layers as triangulated (Fig. 1.42c).

Whereas the right and left ventricular layers (surface) cannot be simultaneously shown with a simple model, the TomTec software can do the LV and RV modeling with numerous triangles. In fact, the related contours to the short-axis view and also connecting the eight points (r_1, r_2, r_3, r_4 and e_1, e_2, e_3, e_4) to their corresponding ones on the epicardium (in each contour) can model the LV and RV layers as triangle (Fig. 1.43a).

The six points between both groups on the contours are protected from being asymmetrical in this model, and these points can be called *controlling points*.

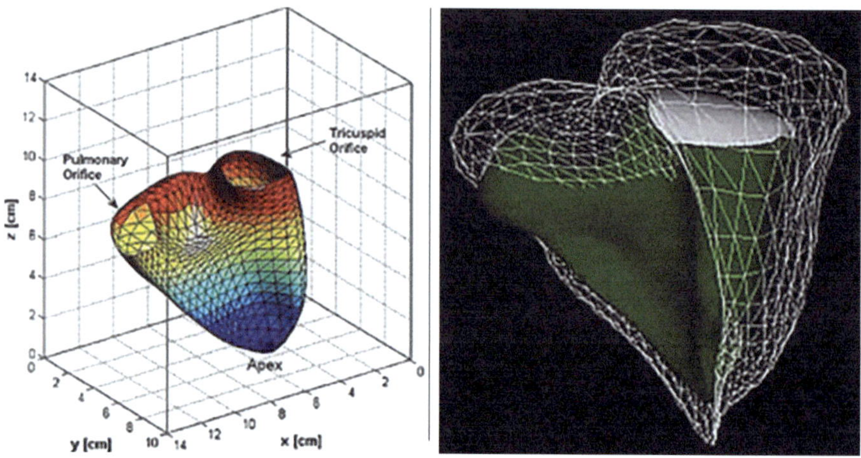

Fig. 1.44 Triangulating the right ventricle in $N_x \times N_Y \times N_z$ using the TomTec software to study the RV in 2D echocardiography systems

Finally, the apex can be reconstructed (Fig. 1.43b), and as this trend continues, both ventricles can be triangulated as well (Fig. 1.43c).

As the layers of triangles are filled (a bright color for the left ventricle and a dark color for the right ventricle), the modeling of these layers completes with 249 triangles and 488 apexes (Fig. 1.43d).

As the number of points increase, the number of triangles increase too. This will lead to better and more exact modeling, and the different ventricular calculations such as deformation, etc., can be more exact [9–12] (Fig. 1.44).

References

1. Burckhardt CB. Speckle in ultrasound b-mode scans. IEEE Trans Son Ultrason. 1978;25:1–6.
2. D'Hooge J, Heimdal A, Jamal F. Regional strain and strain rate measurements by cardiac ultrasound: principles, implementation and limitations. Eur J Echocardiogr. 2000;1:54–70.
3. Jacob G, Noble J, Behrenbruch C, Kelion A, Banning A. A shape-space-based approach to tracking myocardial borders and quantifying regional left-ventricular function applied in echocardiography. IEEE Trans Med Imaging. 2002;21:226–38.
4. Mailloux GE, Langlois F, Simard PY, Bertrand M. Restoration of the velocity field of the heart from two- dimensional echocardiograms. IEEE Trans Med Imaging. 1989;8:143–53.
5. Helle-Valle T, Crosby J, Edvardsen T, Lyseggen E, Amundsen BH, Smith HJ. New noninvasive method for assessment of left ventricular rotation: speckle tracking echocardiography. Circulation. 2005;112:49–56.
6. Dryden IL, Mardia KV. Statistical shape analysis. Chichester: Wiley; 1998.
7. Mikic I, Krucinski S, Thomas JD. Segmentation and tracking in echocardiographic sequences: active contours guided by optical flow estimates. IEEE Trans Med Imaging. 1998;17:274–84.
8. McEachen J, Duncan J. Shape-based tracking of left ventricular wall motion. IEEE Trans Med Imaging. 1997;16:270–83.

9. Kim DH, Kim HK, Kim MK, Chang SA, Kim YJ, Kim MA. Velocity vector imaging in the measurement of left ventricular twist mechanics: head-to-head one way comparison between speckle tracking echocardiography and velocity vector imaging. J Am Soc Echocardiogr. 2009;22(44):52.
10. Kim HK, Sohn DW, Lee SE, Choi SY, Park JS, Kim YJ. Assessment of left ventricular rotation and torsion with two-dimensional speckle tracking echocardiography. J Am Soc Echocardiogr. 2007;20:45–53.
11. Leitman M, Lysyansky P, Sidenko S, Shir V, Peleg E, Binenbaum M. Two dimensional strain—a novel software for real-time quantitative echocardiographic assessment of myocardial function. J Am Soc Echocardiogr. 2004;17:1–9.
12. Kendall DG, Barden D, Carne TK, Le H. Shape and shape theory. New York, NY: Wiley; 1999.

Left Ventricular Modeling as a Polygon

The different models from left ventricular myocardium could help us to generate more studies about the left ventricular's structure and performance.

All of the past researches and studies over the left ventricle have been done based on the cylindrical modeling of the LV myocardium [1]. In this chapter, the LV myocardium would be designed as a polygon. As we know, the rules about a polygon are clear and determined ones; therefore, this subject could help us to have the various calculations on the left ventricle. In fact, a polygon has the different parts as follows: v_i's vertices, e_i's edges and f_i's faces (Fig. 2.1).

In the polygons, the number of vertices is V, the number of edges is E and the number of faces is F. In the nineteenth century, the Swiss mathematician Leonhard Euler was introduced the χ sign as the Euler index for the polygons, and this index would be determined by this formula;

$$\chi = V - E + F;$$

Based on this rule, the Euler index for all simple connected polygons is 2; we can then model the left ventricular myocardium by the same index.

Firstly, the points from apical, mid and basal places would be selected from the left ventricular 3D space; next, these points are connected together, and, finally, the diastolic and systolic images would be achieved [2–6] (Figs. 2.2 and 2.3). Based on Fig. 2.2, $V = 13$, $E = 24$, $F = 13$ and the Euler index is 2.

$$\chi = 13 - 24 + 13 = 2$$

© Springer Nature Switzerland AG 2023

M. Karvandi, S. Ranjbar, *A Review on Recent Echocardiographic Software*,

https://doi.org/10.1007/978-3-031-29046-6_2

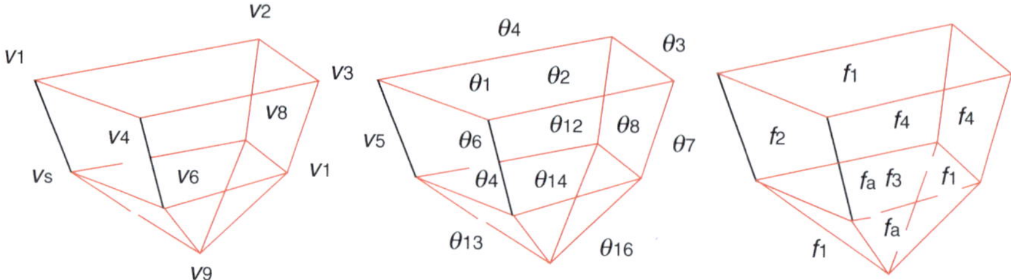

Fig. 2.1 Different parts of a simple connected polygon

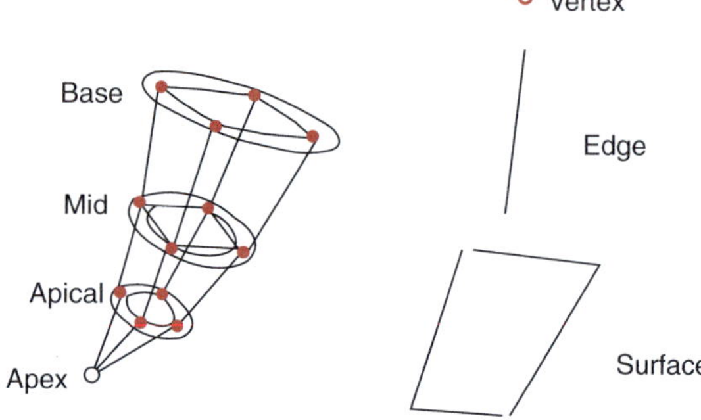

Fig. 2.2 Polygonization of left ventricle and calculating the Euler index

Fig. 2.3 Left ventricle as a polygon

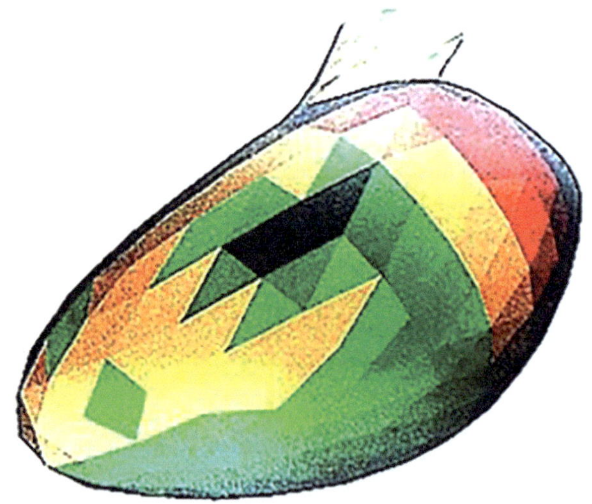

Dehn Number

Two polygons are equivalent that with a similar volume could be altogether changed, as such that by replacing the separated parts from the first polygon, could reach to the second one (Fig. 2.4).

Fig. 2.4 Equivalency of two polygons with similar volume

Fig. 2.5 Necessary parameters in calculating the Dehn number, two faces a and b, edge e and angle $\theta(e)$

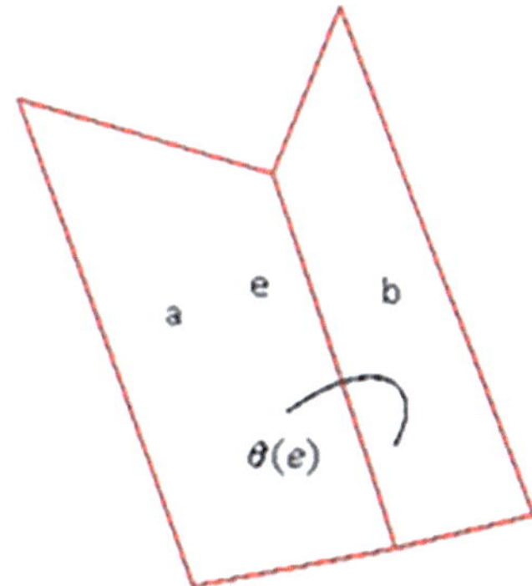

Now, the question is whether all co-volume polygons are equivalent? (This is known as the third problem of Hilbert).

This question was introduced by the German mathematician David Hilbert, and next that was resolved by his student, Max Dehn.

According to the rule of Dehn's invariant number, if two co-volume and simple polygons are equivalent then they have similar Dehn numbers. If we divide the polygon P to the several ones, in fact total Dehn number regarding to the all existed polygons would be similar with the Dehn number of the same polygon P.

Dehn number regarding to the all existed polygons would be similar with the Dehn number of same P polygon. Now, how could we determine the Dehn number in a polygon?

In a polygon, each edge is a denominator of the two adjacent faces, and these faces have a determined angle against each other (Fig. 2.5).

If $l(e)$ is the length of edge and $\theta(e)$ is the angle between the two faces a and b, then the Dehn number is the total aggregated number from the number of the edges' length between the two plates:

$$D(P) = \sum_{e} l(e) \times \theta(e)$$

For instance, in a cubic that has edges of a length of 1 cm, the Dehn number would be as below [7–9] (Fig. 2.6):

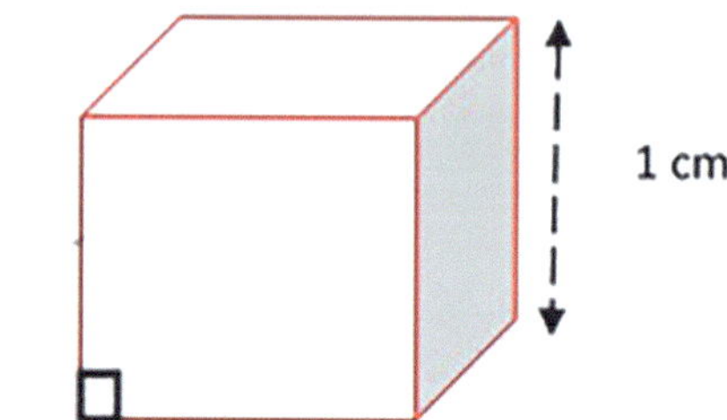

Fig. 2.6 A cubic

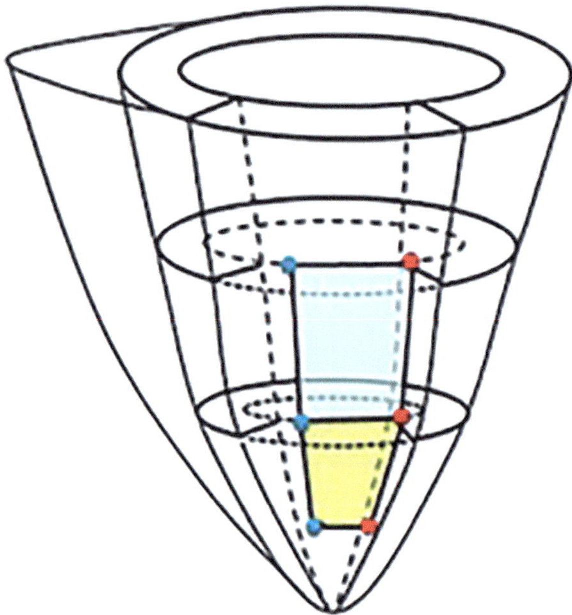

Fig. 2.7 Polygon's vertices, edges and faces of LV; yellow region of apical anterior

$$D(cubic) = 1 \times \frac{\pi}{2} + \ldots 1 \times \frac{\pi}{2} = \frac{13\pi}{2}$$

Some of the sophisticated echocardiography machines would be using from the polygonized LV for the various calculations (Fig. 2.7). In these systems, the left ventricle would be modeled as a polygon with vertices, edges and faces, as if the faces from the interested regions could be shown in the LV. For instance, in Fig. 2.6, the yellow faces is accrued to the anterior apical region as a polygon of LV.

The 2D Speckle tracking software in the LV polygonal model firstly uses each polygonal face as a muscular segment; next, the vertices of this face (muscular segment) would be tracked by the ST program (it should be noted that in the LV cylindrical model, only a point as central segment would be tracked by the ST); finally, the different moving parameters and deformation are calculated (Fig. 2.7).

As seen in Fig. 2.7, the polygon of LV would be meshed in the different regions that each ones are along with a muscular segment of left ventricle. In fact, at any point in the cardiac cycle, a polygon would be built that would have the Dehn number. This number is fixed one in the different phases Iso-volumic relaxation (IVR) and Iso-volumic contraction (IVC), since there is no change at all in them.

In Fig. 2.7 (mid image), the yellow region with the four red vertices could show the basal anterior in the end-diastole phase, and (right picture) the green region with the four red vertices could show the same segment in the end-diastole phase (contrary with the LV cylindrical model that only one point would be taken as central segment). In Fig. 2.8, the left ventricular polygonization would be shown in a single cardiac cycle; the main subject is calculating the vertices' distance and also their displacements in the polygon has not been as a regular and linear Euclidean calculation via the LV cylindrical model, but it has been generated as a logarithmical model (Fig. 2.9).

For instance, if $\lambda(u)$ is one of the polygonal vertices in Fig. 2.8 (right) and α_Λ is the other vertex of the same sample, their distance would be calculated as below:

$$\rho\left(\lambda(u),\alpha_\Lambda\right) = \begin{cases} -\log|u|_{ir} & \text{for}\,|u| \leq 1 \\ -\log\left|\dfrac{1}{u}\right|_{ir} & \text{for}\,|u| \geq 1. \end{cases}$$

In this formula, q is a reliable number and is also one of the polygonal indexes, and is a square from 1. Those polygons would be used for modeling of LV; $q = 2$ or $q = 3$. In Fig. 2.10, the movement of a vertex toward the adjacent vertices is shown.

The motion of each mesh (network) and also its vertices in a cardiac cycle would be considered by the logarithmical model. The q index could be controlled the quality of a vertex's adjacency with the other vertices, and also its finding direction during time-based moving. In fact, the motion of a polygon of LV would start from a vertex and could reach the adjacent vertices under the control of q, then the first mesh (network) would be built in the next frame (Fig. 2.10). According to Fig. 2.11, after being built of the first mesh (network), the closest vertex to the first mesh would be determined by the logarithmical model, and also, based on the same q, the next meshes (networks) would be built in the next frames; therefore, the dynamic motion of this polygon can be created in a cardiac cycle (Fig. 2.11).

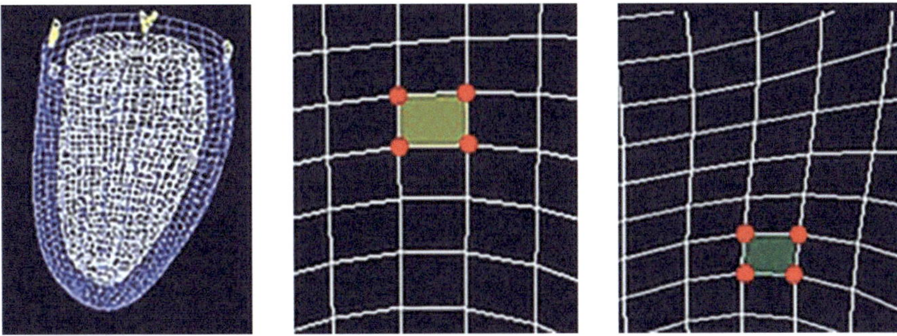

Fig. 2.8 LV polygonal modeling (left), assigning the face, vertex and its tracking in end-diastole (mid) and end-systole phase (right)

Constructing the polygon of LV as a static form in a cardiac cycle (left panel in Fig. 2.8) and considering the motion and deformation of each apex and polygonal plates by the logarithmical metric (Figs. 2.10 and 2.11) could lead us toward software that can be posted on the echocardiography machines as well [9–11] (Fig. 2.12).

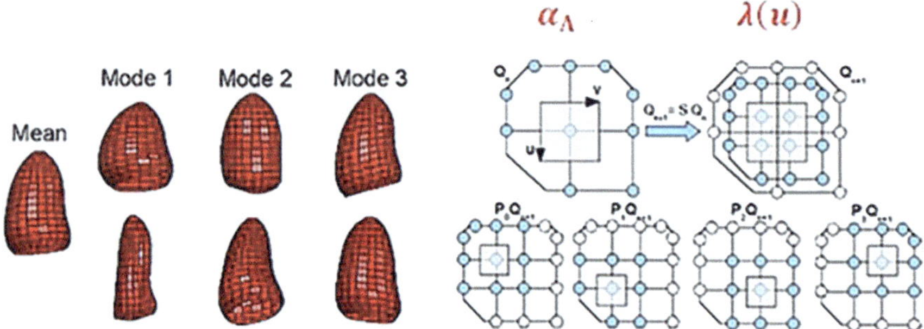

Fig. 2.9 Polygon of LV in different phases of a cardiac cycle (left), meshing the different regions of LV, motion path from a polygonal vertex to other one, and meshes' distance (networks) based on a logarithmical metric (right)

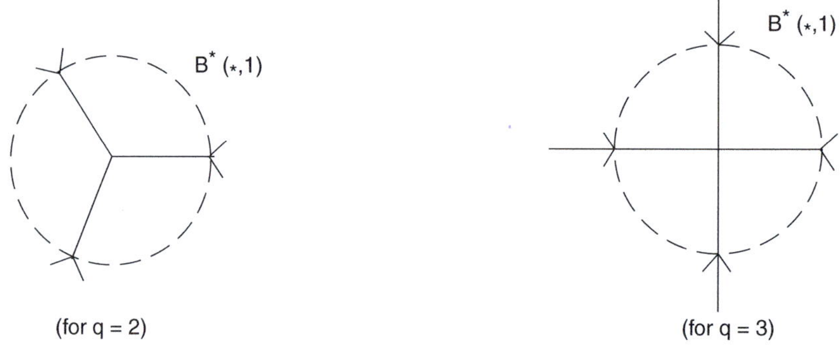

Fig. 2.10 Starting point of a vertex's moving to other vertices and building a mesh (network) in adjacency as 1 radius; $q = 2$ (left) and $q = 3$ (right)

Fig. 2.11 Built meshes (networks) based on logarithmical model; q is 2 and forming a dynamical motion related to the polygon of LV in a cardiac cycle

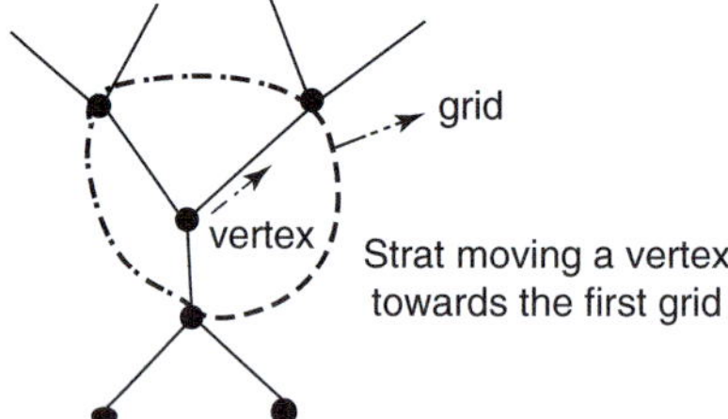

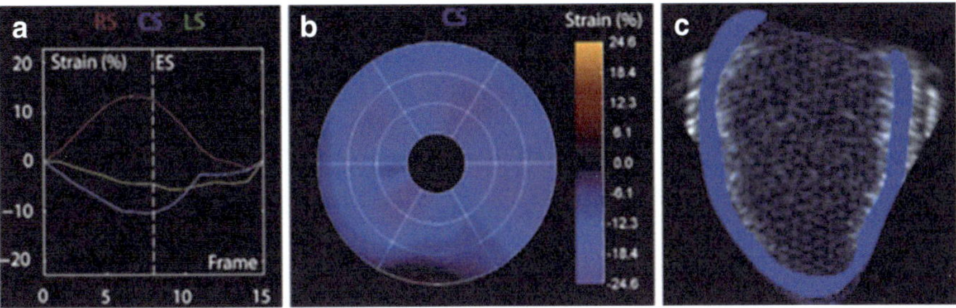

Fig. 2.12 (a–c) Using the polygonization software of LV in 2D echocardiography systems and 4D strain measurements

References

1. Mor-Avi V, Lang RM, Badano LP, Belohlavek M, Cardim NM, Derumeaux G. Current and evolving echocardiographic techniques for the quantitative evaluation of cardiac mechanics: ASE/EAE consensus statement on methodology and indications endorsed by the Japanese Society of Echocardiography. Eur J Echocardiogr. 2011;12:167–205.
2. Rogers M, Graham J. Robust active shape model search. In: Proceedings of the European Conference on Computer Vision; 2002. p. 517–30.
3. Suhling M, Arigovindan M, Hunziker P, Unser M. Motion analysis of echocardiograms using a local-affine, spatio-temporal model. In: Proceedings of the IEEE International Symposium on Biomedical Imaging: from macro to nano; 2002. p. 573–6.
4. Paragios N. A variational approach for the segmentation of the left ventricle in MR cardiac images. In: Proceedings IEEE workshop on variational and level set methods; 2001.
5. Otto CM. The practice of clinical echocardiography. Philadelphia, PA: Saundes Elsevier; 2011.
6. Pirat B, Khoury DS, Hartley CJ, Tiller L, Rao L, Schulz DG. A novel feature tracking echocardiographic method for the quantitation of regional myocardial function: validation in an animal model of ischemia- reperfusion. J Am Coll Cardiol. 2008;51:1–9.
7. Clarysse P, Friboulet D, Magnin IE. Tracking geometrical descriptors on 3-D deformable surfaces: application to the left-ventricular surface of the heart. IEEE Trans Med Imaging. 1997;16:392–404.
8. Truesdell CA. A critical summary of developments in nonlinear elasticity. J Ration Mech. 1953;1:125–300.
9. Truesdell CA. A critical summary of developments in nonlinear elasticity. J Ration Mech. 1953;2:593–616.
10. Torrent-Guasp F, Kocica MJ, Corno A, Komeda M, Cox J, Flotats A, Ballester-Rodes M, Carreras-Costa F. Systolic ventricular filling. Eur J Cardiothorac Surg. 2004;25:376–86.
11. Cootes T, Taylor C. Active shape models-'Smart Snakes'. In: Proceedings British Machine Vision Conference; 1992. p. 266–75.

Left Ventricular Torsion Based on Echocardiographic Data

3

The torsion of the left ventricular muscle due to helical motion in the muscular fibers, in fact, is one of the most important parameters to having studied left ventricular performance, and it could be achieved from the total difference of LV's rotation on the basal and apical layers. This chapter is aimed to consider and calculate the torsion in the left ventricle via the new and modern strain tracking technology (ST) in the echocardiography systems. Firstly, the motion and interested points' regions on the left ventricular basal and apical layers and in a cardiac cycle was tracked by the same ST as the frame to frame trend. Then, all images in the six separated and consistent frames in the basal and apical layers and in the short-axis views are taken from the end-diastole to the end-systole phase. According to Fig. 3.1, the red points could be determined the epicardial and endocardial borders, and based on these the middle of the myocardial muscle, or the Region of Interest (ROI), would be brightly determined. The center of left ventricular cavity is a red point that also has a small cross in the center of the same cavity, and the blue cross lines in each frame could indicate the rotational average of LV (Fig. 3.1).

In considering the left ventricular torsion, firstly the images should be taken from the basal and apical layers and in the short-axis views, then the anterior, posterior, septum and lateral segments would be determined by the ST software. According to Fig. 3.2, the velocity difference in the two opposite segment septum and lateral segments would be calculated as $v_{lat}(t) - v_{sep}(t)$, and then the area between the velocity curves ($R(t)$) in the anterior and posterior segments would also be determined. Figures 3.2 and 3.3 can show the angular motion of left ventricle as the same measurement. So, the left ventricular rotational velocity at the basal and apical layers would be calculated and then it would be shown as follows; $LV_{rot,\,v}(t)$ (Figs. 3.2 and 3.3).

Calculating the rotational value of LV in the different segments in a cardiac cycle is done as below:

© Springer Nature Switzerland AG 2023

M. Karvandi, S. Ranjbar, *A Review on Recent Echocardiographic Software*,

https://doi.org/10.1007/978-3-031-29046-6_3

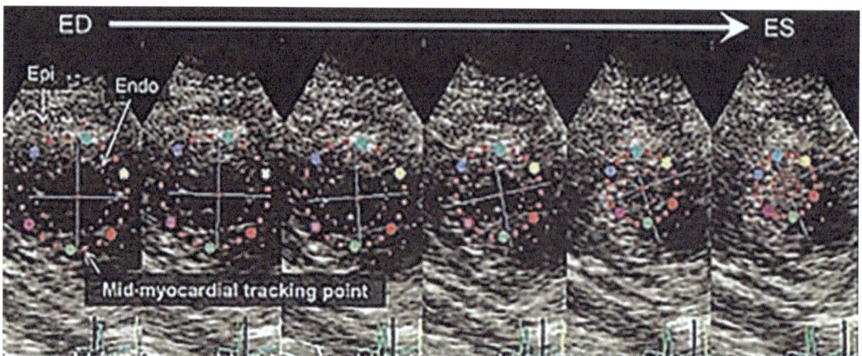

Fig. 3.1 Approximating stages of left ventricular torsion at basal and apical layers in a cardiac cycle by Speckle tracking software

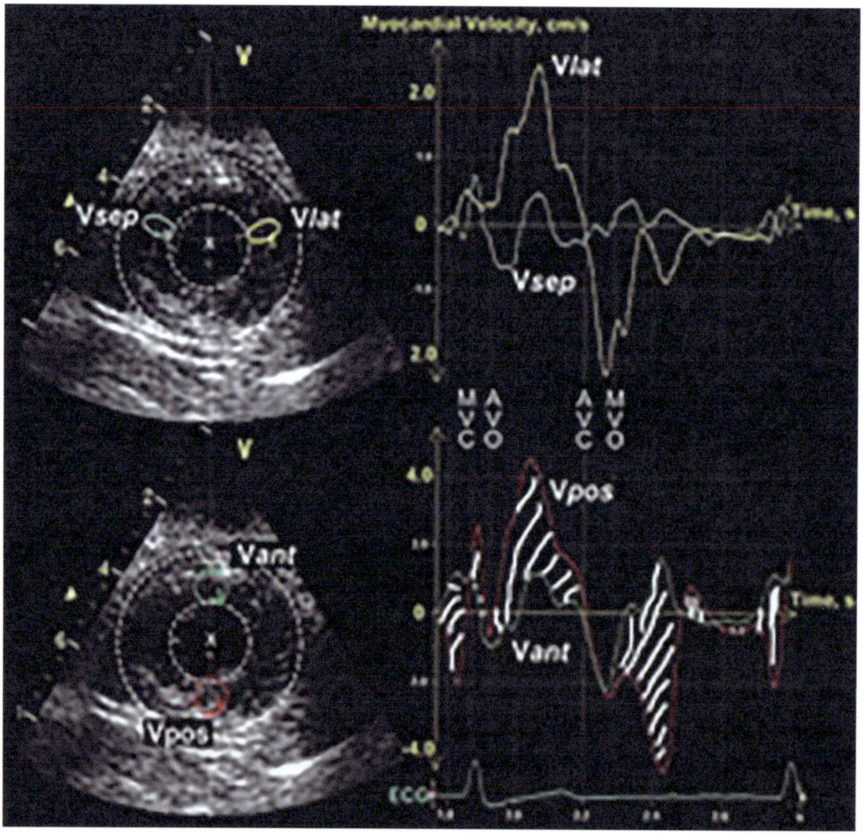

Fig. 3.2 Dividing the short-axis view to anterior, posterior, lateral and septum regions and their velocity curves; the hatched region is the measurement between velocity curves of anterior and posterior segments ($R(t)$)

Fig. 3.3 Angular motion of a segment

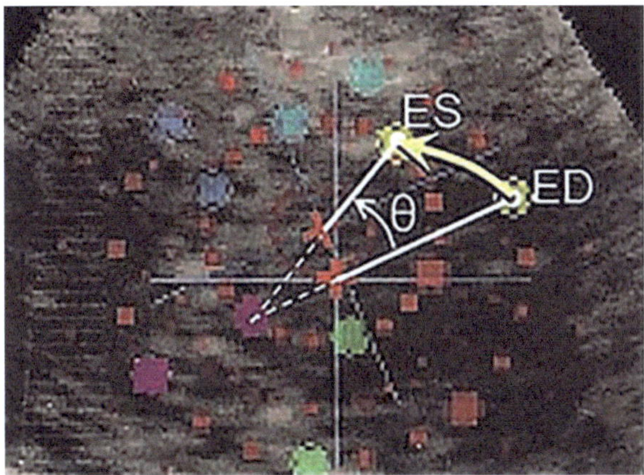

$$LV_{rot,v}(t) = \frac{\left[v_{sep}(t) - v_{lat}(t)\right]}{2R(t)};$$ (3.1)

$$R(t) = \frac{R_0}{2} + \frac{1}{2}\int_0^{t_s}\left(v_{ant}(t) - v_{pos}(t)\right)dt$$ (3.2)

Then,

$$LV_{rot,v}(t) = \frac{\left[v_{sep}(t) - v_{lat}(t)\right]}{2\left[\dfrac{R_0}{2} + \dfrac{1}{2}\int_0^{t_s}\left(v_{ant}(t) - v_{pos}(t)\right)dt\right]}$$ (3.3)

In fact, R_0 is the left ventricular inner radius in the end-diastole phase and t_s is the time in the end-systole phase.

By including formula (3.3), calculating the left ventricular rotational velocity in each section and any different phase would be possibldone to achieve. The related integral could determine the left ventricular rotational velocity based on the angle in basal and apical sections, and introduce the formula below [1]:

$$LV_{tor} = \int_0^{t_s} LV_{tor,v}\,dt$$

In a normal heart, the left ventricular rotational velocity at the basal and end-systole phase is $8\pm1.6°$ clockwise, and at the apical and end-systole phase is $9.5\pm1.8°$ counterclockwise. It is worthwhile mentioning that the direction of this rotation would be determined by the ST software (Fig. 3.4).

The left ventricular torsion value (LV torsion) could be achieved by the difference of the same torsion value at the apical and basal segments, and this is $17.3\pm2.5°$ (Fig. 3.5).

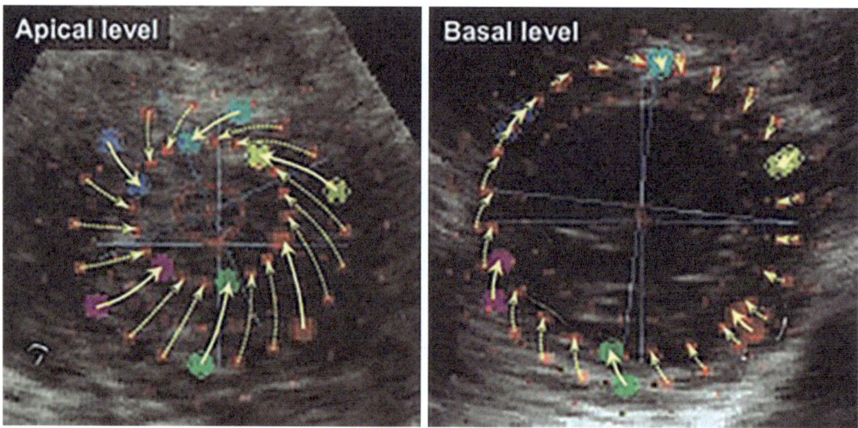

Fig. 3.4 Direction of left ventricular rotation at basal and apical segments in end-systole phase and start-flashes shows end-diastole and end-systole phases respectively

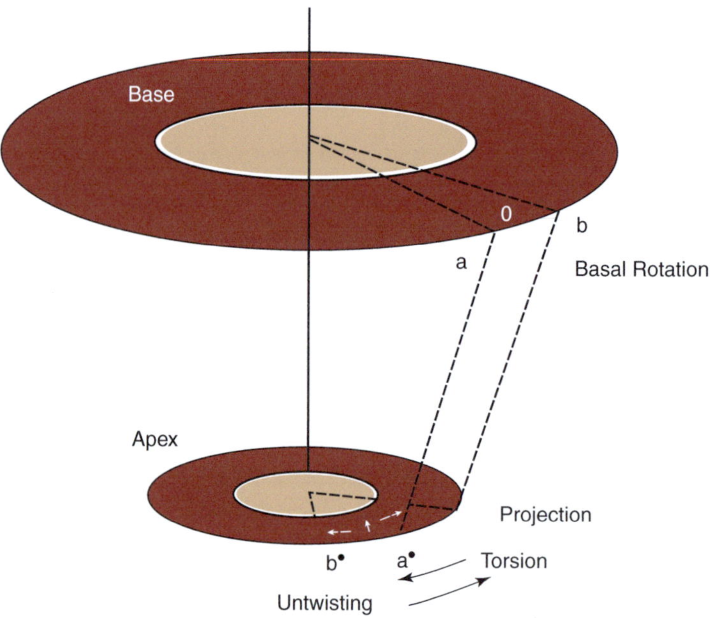

Fig. 3.5 LV torsion and rotation at apical and basal segments

The rotation direction of LV at the orthogonal parts of circle in a cardiac cycle—assuming the LV cylindrical form—could be achieved by the ST software (Fig. 3.6). It is obvious that the numerical values of the shown rotational parts could be calculated by the mentioned formulas.

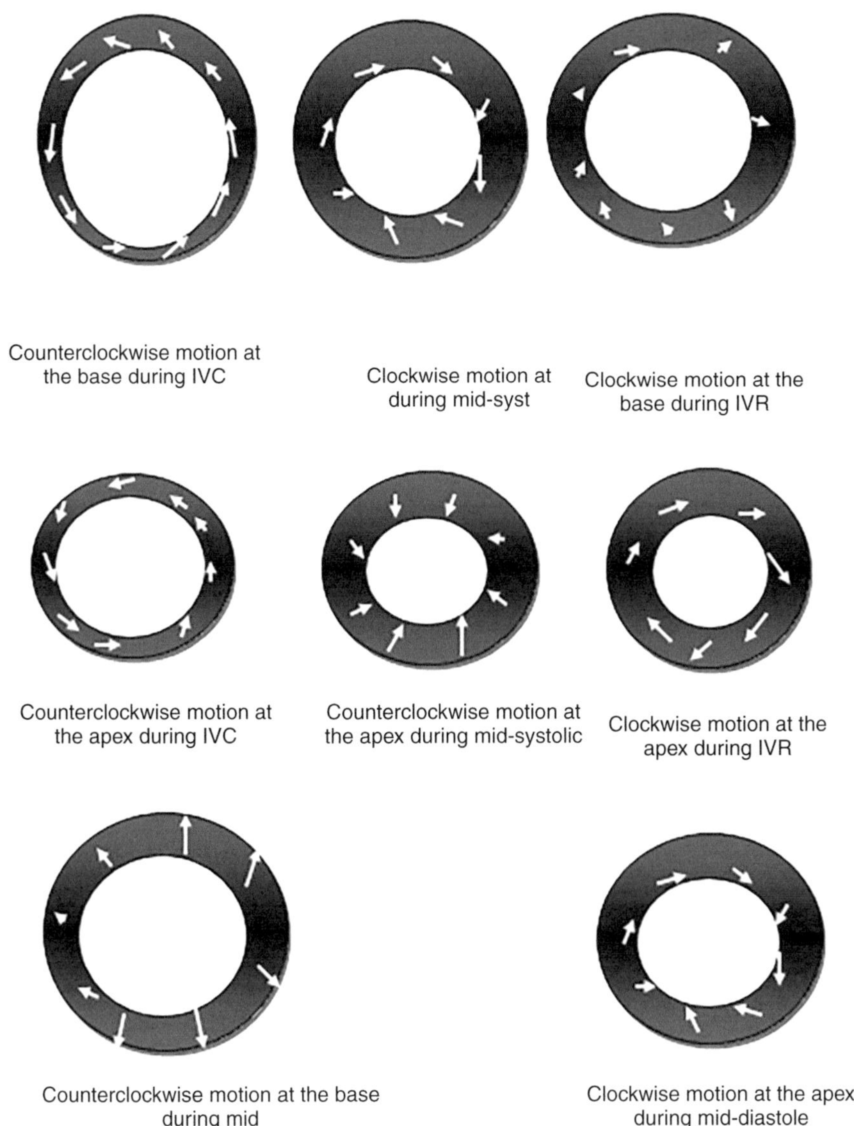

Fig. 3.6 Direction of muscular fibers' rotational motions from base to apex in different phases

In the sophisticated echocardiography systems, considering the rotation of LV is based on a 4D modeling of LV. As stated in Chap. 2, LV could be modeled as apolygon (Fig. 3.7); in fact, the motion of its faces, vertices and edges in a cardiac cycle could be determined the same 4D model of LV. After being built from the dynamic model of LV, we can calculate the rotation of LV by the 2D SP software [2, 3].

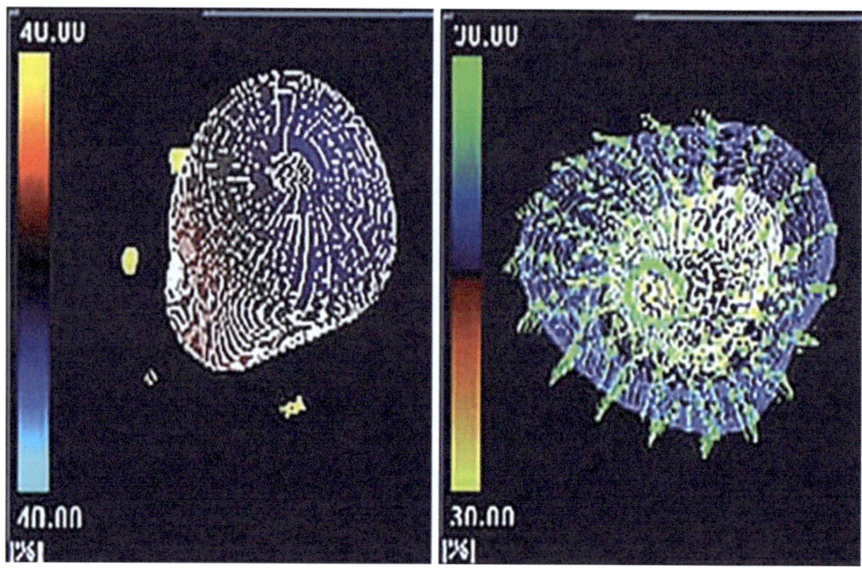

Fig. 3.7 Polygonal 4D model of LV and its rotation in apical and basal segments

References

1. Kim HK, Sohn DW, Lee SE, Choi SY, Park JS, Kim YJ. Assessment of left ventricular rotation and torsion with two-dimensional speckle tracking echocardiography. J Am Soc Echocardiogr. 2007;20:45–53.
2. Otto CM. The practice of clinical echocardiography. Philadelphia, PA: Saundes Elsevier; 2011.
3. Zhou XS, Comaniciu D, Krishnan S. Coupled-contour tracking through non orthogonal projections and fusion for echocardiography. In: Proceedings of the European Conference on Computer Vision; 2004.

Motion Analysis of Left Ventricular Inner Wall

4

Connection between the right and left ventricle is like an interconnected hank. This chapter is mainly aimed to present the new methods, so that we can find their complexities and also can reach to a range of the more comprehensive models about the moving direction of cardiac muscular fibers, and also the different methods in tracking the left ventricular muscular border. The spiral motion of the same muscular fibers from the base to apex would be viewed as below [1–3] (Fig. 4.1).

In Fig. 4.2, the arrangement quality of these spiral muscular fibers and their motion from the epicardium to endocardium has been shown.

In this chapter, the three analytical methods for the left ventricular motion in a cardiac cycle would be presented [4].

1. Closest point method.
2. Centerline method.
3. Space border method.

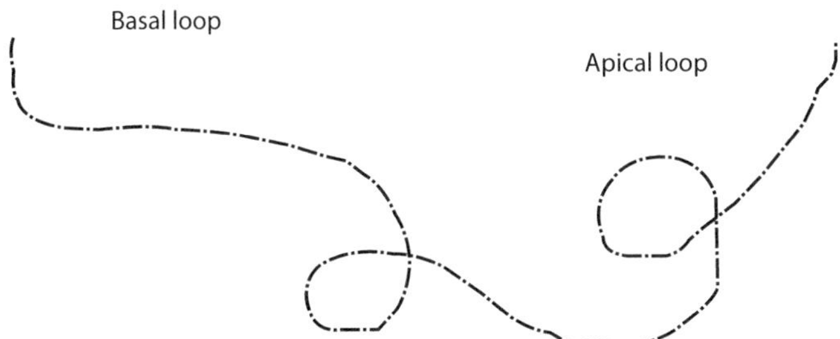

Fig. 4.1 A schematic view from spiral motion of a muscular fiber

© Springer Nature Switzerland AG 2023

M. Karvandi, S. Ranjbar, *A Review on Recent Echocardiographic Software*,

https://doi.org/10.1007/978-3-031-29046-6_4

Closest Point Method

End-diastolic border (EDB) is the left ventricular inner border or edge in the end-diastole phase and end-systolic border (ESB) is the left ventricular inner border or edge in this phase (Fig. 4.3).

In this method, we could pointing on the EDB with interested numbers, then from the closest distance with regard to each point on the EDB up to its corresponding point on the ESB would be determined; finally, the corresponding lines should be drawn (Figs. 4.4 and 4.5).

Now, the question is: how should the closest point be selected?

Fig. 4.2 Moving direction of muscular fibers from epicardium to endocardium

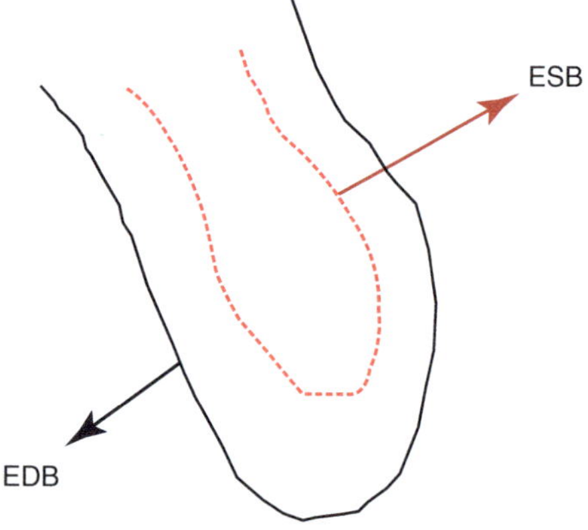

Fig. 4.3 Left ventricular inner border in end-diastolic phase (black curve), and left ventricular inner border in end-systolic phase (red strip line)

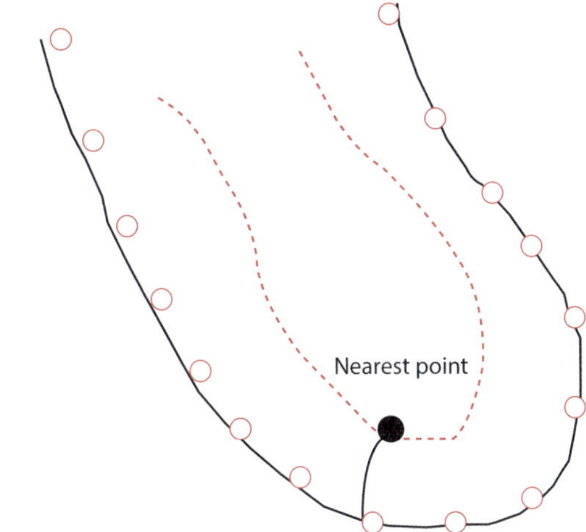

Fig. 4.4 Closest point on the left ventricular inner edge in the end-systole phase up to the targeted point on the left ventricular inner edge in the end-diastole phase

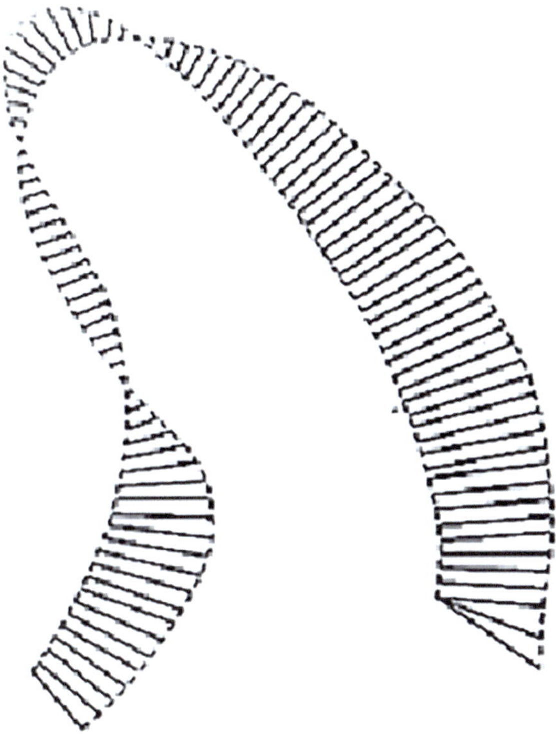

Fig. 4.5 Connecting line segments of the closest points on left ventricular inner edge in end-systole and diastole phases

If it is assumed that "*x*" is a selected point on the inner edge in the EDB, then there is a "*y*" point on the ESB that is the closest point when compared with the other points of the ESB to the *x* on the EDB.

Now, how would "*y*" be found on the ESB? Since the ESB is a closed curve and also that *x* is not accrued to the ESB, the least distance of *x* on the ESB would be calculated as below;

$$d = \min\left\{x - z; z \in \mathrm{ESB}\right\};$$

(*z* is all interested points on the ESB).

If we draw a range of the circles with different radius as $d + \dfrac{1}{m}$ with the centrality of *x* on the EDB, such that all circles would have crossed the ESB, then the smallest radius shows the least distance (*d*). In fact, the joint of these circles is point *y* (joint theorem of contour segments),

$$\mathrm{ESB}_k = \left\{z \in \mathrm{ESB}; x - z \le d + \frac{1}{M}\right\}$$

$$\mathrm{ESB}_1 \supseteq \mathrm{ESB}_2 \supseteq \dots \supseteq \mathrm{ESB}_k \supseteq \dots$$

In fact, central circles are *x*, and their radius is $d + \dfrac{1}{k}$, and joint theorem is $\cap\, \mathrm{ESB}'_k\mathrm{s} = \varnothing$ (Figs. 4.6 and 4.7).

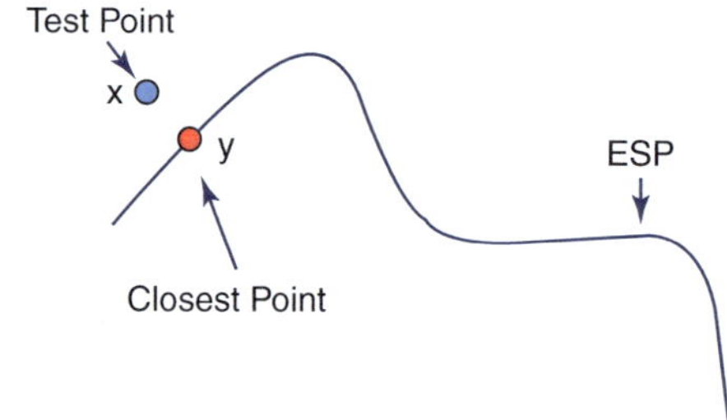

Fig. 4.6 Closest point (*y*) on the ESB to the targeted point (*x*)

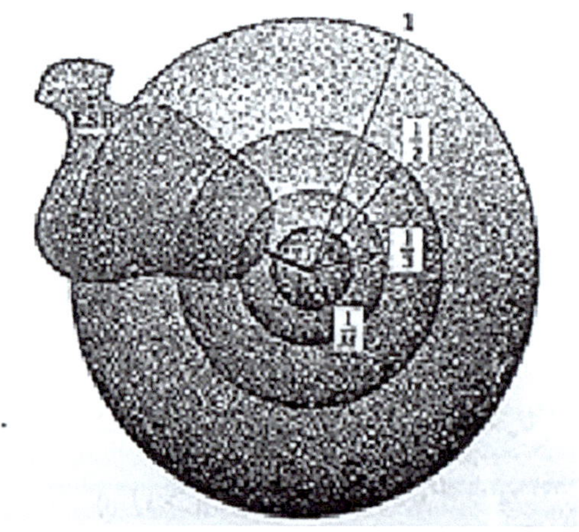

Fig. 4.7 Determining trend of closest point against targeted point by circles

Utility of Closest Point Method

The drawn corresponding lines have many medical conceptions. Figure 4.7 is an animated figure from the corresponding lines, and it can also indicate any change in these lines' length per a single cardiac cycle. Therefore, the length shortness and slow motion of these lines in a cardiac cycle is known as an abnormal sign; for instance, a segment that has hyperkinesia, its corresponding lines' length and its motion would decrease. Hence, when the lines' length decreases or increases, or when the lines' velocity decreases or increases, we can provide an exact interpretation about the motion of the cardiac muscle.

Centerline Method

Centerline method for regional wall motion analysis study end diastolic and end systolic endocardial contours of the left ventricle from the apical four chamber echocardiographic view. One hundred equidistance chords are constructed perpendicular to the centerline between the end-diastolic and end-systolic endocardial contours. The motion at each chord is normalized by the end diastolic perimeter to yield a dimensionless chord fractional lengthening and shortening.

Space Border Method

Firstly, we assume a point as x on the ESB, whereas ESB is accrued to the EDB, then there is a circle with x as its center and radius as $0 < \delta(x)$ such that it does not cross the EDB's scope. In fact, $\delta(x)$ is the closest point between x on the ESB and its corresponding point on the EDB.

Now, for any points x on the ESB, we could be shown a circle with x as its center and a radius as $\delta(x)/2$ $(D(x))$. It is obvious that all circles $D(x)$ could also cover the whole of the ESB.

Because the ESB is a limited, closed and bordering curve, therefore based on the Heine-Borel Theorem, the finite number of $(x_1, x_2, \ldots, x_n)$ points would be existed on the ESB, and their aggregation, including $D(x_1)$, $D(x_2)$, $\ldots$, $D(x_n)$, could be covered the whole of the ESB [5–8] (Fig. 4.8).

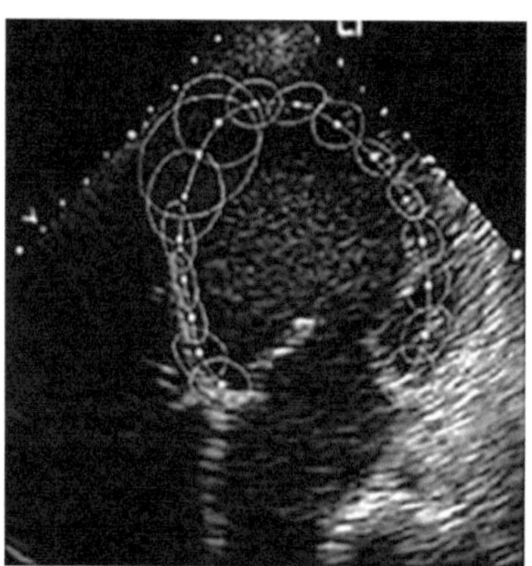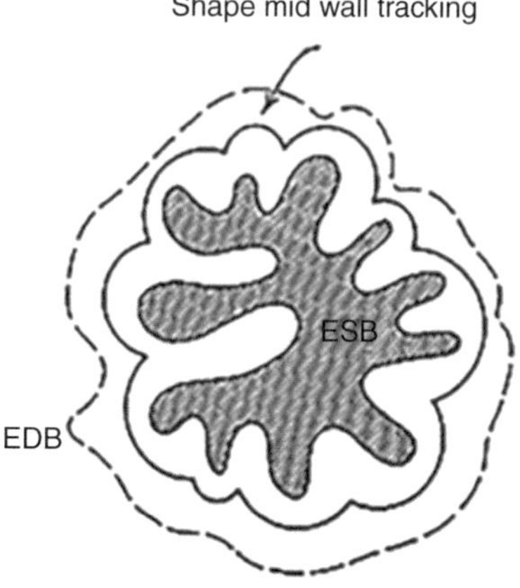

Fig. 4.8 Points and their corresponding circles (left) and ESB coverage by achieved circles (right)

Utility of the Space Border Method

This method, as stated above, is a new method in the shape tracking of the left ventricular myocardial muscle, and border tracking methods could be followed the lines and points of the myocardial muscle's border (Fig. 4.9).

In this new method, the space border (edge) of an interested point would be tracked from a frame to the other one on the circle's bow (Fig. 4.10).

Since this method could be tracked the myocardial muscle's border by the circle shapes, then increasing and decreasing the radius of these circles could be led to the exact description. For instance in LV enlargement, from the apex to the base, these circles will shrink, or, in other words, the circles' radius will decrease [9, 10] (Fig. 4.11).

Fig. 4.9 Tracking the left ventricular inner edges by border tracking software

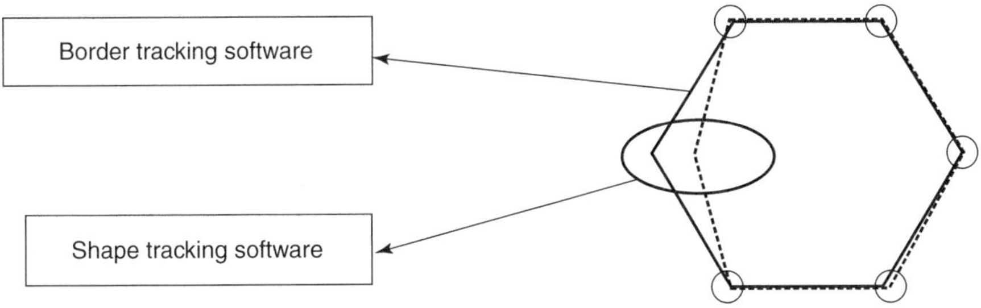

Fig. 4.10 Moving on circle's bow in the shape tracking method

Fig. 4.11 Decrease in the circles' radius from the left ventricular apex to its base in LV enlargement—shape tracking software

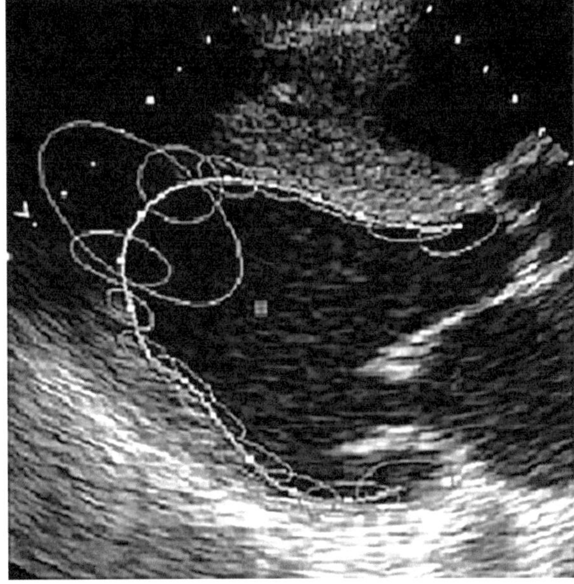

References

1. Burckhardt CB. Speckle in ultrasound b-mode scans. IEEE Trans Son Ultrason. 1978;25:1–6.
2. D'Hooge J, Heimdal A, Jamal F. Regional strain and strain rate measurements by cardiac ultrasound: principles, implementation and limitations. Eur J Echocardiogr. 2000;1:54–70.
3. Kendall DG, Barden D, Carne TK, Le H. Shape and shape theory. New York, NY: Wiley; 1999.
4. Otto CM. The practice of clinical echocardiography. Philadelphia, PA: Saundes Elsevier; 2011.
5. Dryden IL, Mardia KV. Statistical shape analysis. Chichester: Wiley; 1998.
6. Mikic I, Krucinski S, Thomas JD. Segmentation and tracking in echocardiographic sequences: active contours guided by optical flow estimates. IEEE Trans Med Imaging. 1998;17:274–84.
7. McEachen J, Duncan J. Shape-based tracking of left ventricular wall motion. IEEE Trans Med Imaging. 1997;16:270–83.
8. Kim DH, Kim HK, Kim MK, Chang SA, Kim YJ, Kim MA. Velocity vector imaging in the measurement of left ventricular twist mechanics: head-to-head one way comparison between speckle tracking echocardiography and velocity vector imaging. J Am Soc Echocardiogr. 2009;22:44–52.
9. Kim HK, Sohn DW, Lee SE, Choi SY, Park JS, Kim YJ. Assessment of left ventricular rotation and torsion with two-dimensional speckle tracking echocardiography. J Am Soc Echocardiogr. 2007;20:45–53.
10. Leitman M, Lysyansky P, Sidenko S, Shir V, Peleg E, Binenbaum M. Two dimensional strain—a novel software for real-time quantitative echocardiographic assessment of myocardial function. J Am Soc Echocardiogr. 2004;17:1–9.

Generalized Strain Components of the Left Ventricular Myocardium

5

It is essential to understand the physiology of the LV function in order to interpret deformation imaging. Normal LV systolic function is the result of coordinated contraction of myocardial fibers of different orientations. During systole, the longitudinal component of myocardial contraction causes the LV base to descend approximately 12–15 mm toward the apex, while the circumferential component causes a reduction in the LV short-axis diameter, with both mechanisms contributing to a wall thickening of approximately 50%. In addition, due to the contraction of obliquely oriented fibers, the LV undergoes a twisting or wringing deformation along its long axis, involving oppositely directed rotations of the apex relative to the base (Fig. 5.1). Systolic LV deformation is reversed or reformatted during diastole, primarily during the early phase.

Representative strain and strain rate traces of normal and ischemic myocardium during a cardiac cycle are shown in Fig. 5.2. In normal ventricles, the systolic deformation of an elastic myocardium contributes to the early diastolic lengthening or reversal of systolic deformation which is a consequence of the rate of active myocardial relaxation, the release of restoring forces generated during systolic deformation, as well as the load applied to the LV during early diastole. Passive-elastic properties and the pericardium modulate myocardial deformation. Magnitude and rate of systolic and diastolic deformation along the three main LV axes (longitudinal, circumferential, and radial) can be quantified by measurements of strain and strain rate, respectively. The more complex wringing LV deformation can be assessed by twist and torsion [1–3].

The assessment of the heart's ventricular function is the most frequent indication for echocardiography. Heart ventricular systolic and diastolic dysfunction are reliable indicators of mortality.

The left ventricular wall is composed of both circular and longitudinal myocardial fibers. In pathologic conditions (e.g. myocardial ischemia or hypertrophy), long-axis myocardial function is impaired first. Thus, this indicates an early diagnosis of cardiac dysfunction.

© Springer Nature Switzerland AG 2023

M. Karvandi, S. Ranjbar, *A Review on Recent Echocardiographic Software*,

https://doi.org/10.1007/978-3-031-29046-6_5

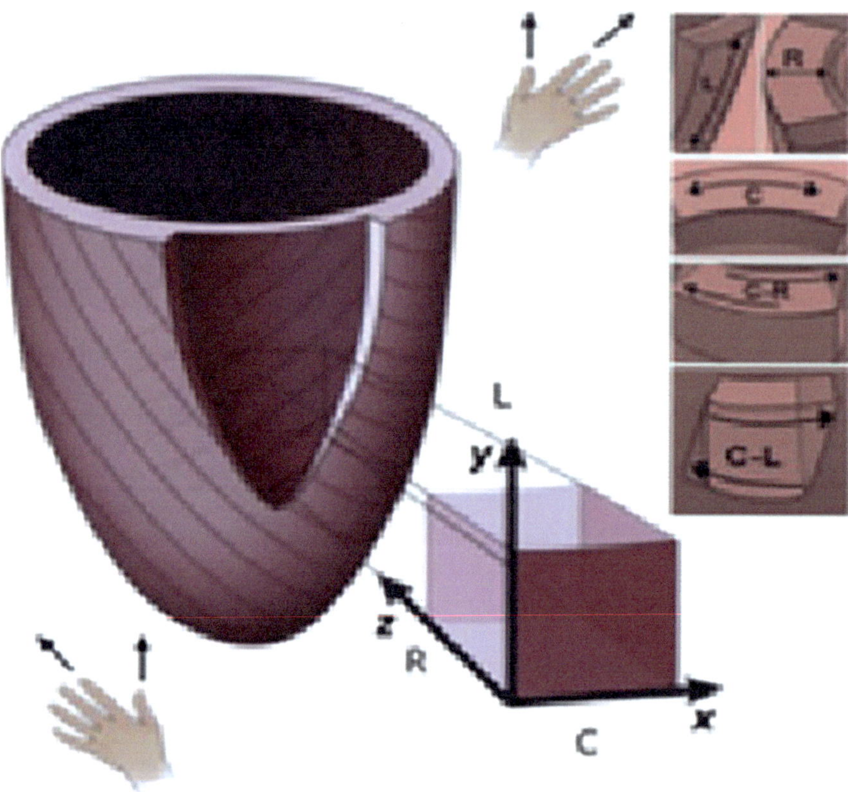

Fig. 5.1 Left ventricular fiber orientation and three-dimensional deformation. The left panel shows a schematic representation of the myocardial fiber orientation in the left ventricle that changes continuously from a right-handed helix in the subendocardial region to a left-handed helix in the subepicardial region, as seen over the anterior wall of the left ventricle. The panels to the right show block of myocardial tissue in which the x-axis is oriented at a tangent to the circumferential (C) direction, the y-axis is oriented longitudinally (L), and the z-axis corresponds to the radial (R) direction of the left ventricle and the components of shear strain (CR, CL)

Characterization of local and global contractile activities in the myocardium is essential for a better understanding of cardiac form and function. The spatial distribution of regions that contribute the most to cardiac function plays an important role in defining the pumping parameters of the myocardium like ejection fraction and dynamic aspects such as twisting and untwisting.

In general, myocardial shortening and lengthening, tangent to the wall, and ventricular wall thickening are important parameters that characterize the regional contribution within the myocardium to the global function of the heart.

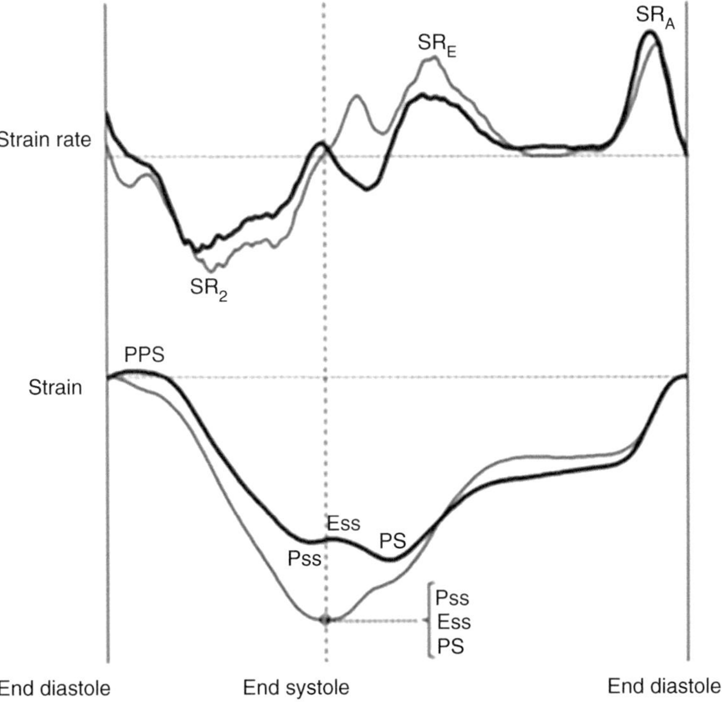

Fig. 5.2 Strain and strain rate from normal and ischemic myocardium. Strain rate and strain traces from typical normal (gray) and mildly ischemic (black) myocardium. Ischemic myocardium is characterized by an early peak positive strain (PPS), and peak systolic strain (PSS) is typically lower than endsystolic strain (ESS). A post-systolic shortening is often seen, and peak strain (PS) therefore occurs after end systole

Motion

- Displacement (mm) = velocity × time.
- Velocity.

Deformation

- Strain (1-D, 2-D and 3-D).
- Strain rate (1-D, 2-D and 3-D).

1D Strain

1D strain of a cardiac segment means its shortening or lengthening within a cardiac cycle and the rate of strain would be strain rate. On the other hand, strain rate evaluates the velocity of the shortening and lengthening (Fig. 5.3).

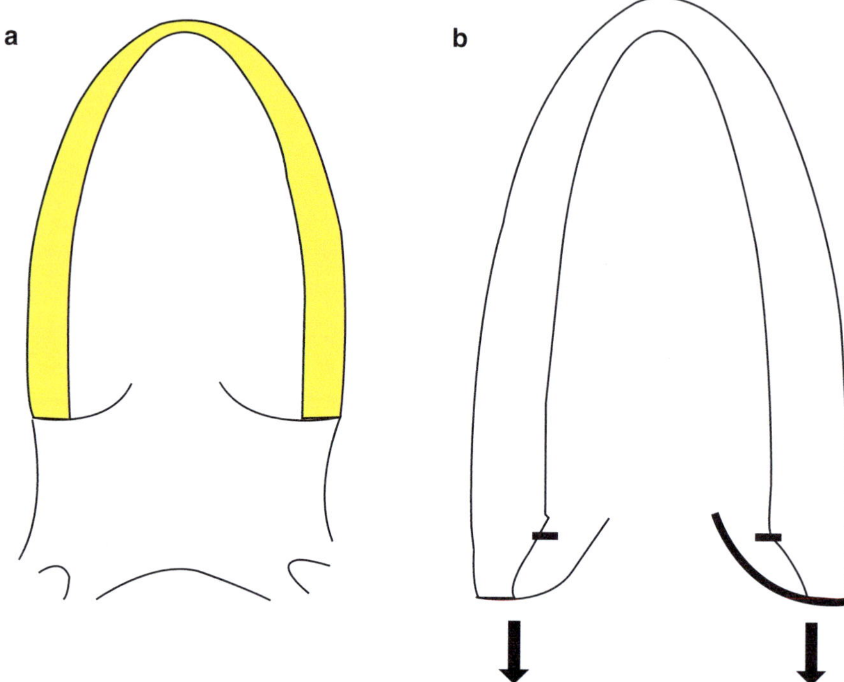

Fig. 5.3 Left ventricular myocardial lengthening longitudinally (**a**, **b**)

2D Strain

2D strain of a cardiac muscle sample is a two-by-two matrix that describes all its deformation to radial and longitudinal directions (Figs. 5.4 and 5.5).

These 2D modifications can be represented as the following matrix:

$$\begin{pmatrix} \dfrac{\Delta x}{x} & \dfrac{\Delta x}{y} \\[2mm] \dfrac{\Delta y}{x} & \dfrac{\Delta y}{y} \end{pmatrix} \text{Such that,} \ \frac{\Delta x}{y} = \tan\theta_y, \ \frac{\Delta y}{x} = \tan\theta_x$$

We call these entries as the 2D strain components by $e_{11} = \dfrac{\Delta x}{x}$, $e_{12} = \dfrac{\Delta x}{y}$, $e_{21} = \dfrac{\Delta y}{x}$ and $e_{22} = \dfrac{\Delta y}{y}$.

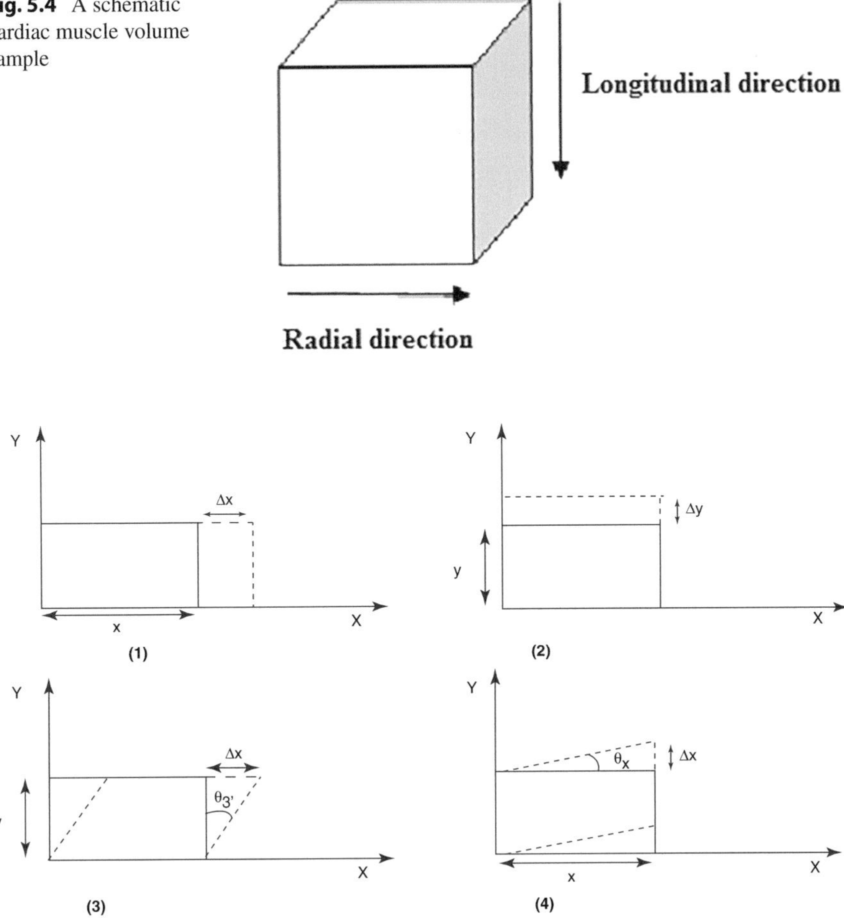

Fig. 5.4 A schematic cardiac muscle volume sample

Fig. 5.5 2D changes of a cardiac muscle sample respect to the x–y axis

3D Strain

By a similar argument we can give an explicit description of the 3D strain of a myocardial muscle volume sample (Fig. 5.6).

3D strain of a cardiac muscle volume sample is a three-by-three matrix that states all its deformation to radial, longitudinal and circumferential directions (Fig. 5.7).

These 3D changes can be represented as the following matrix:

$$\begin{pmatrix} \varepsilon_x & \varepsilon_{xy} & \varepsilon_{xz} \\ \varepsilon_{yx} & \varepsilon_y & \varepsilon_{yz} \\ \varepsilon_{zx} & \varepsilon_{zy} & \varepsilon_z \end{pmatrix},$$

Fig. 5.6 Attached a system of coordinates to a cardiac muscle volume sample

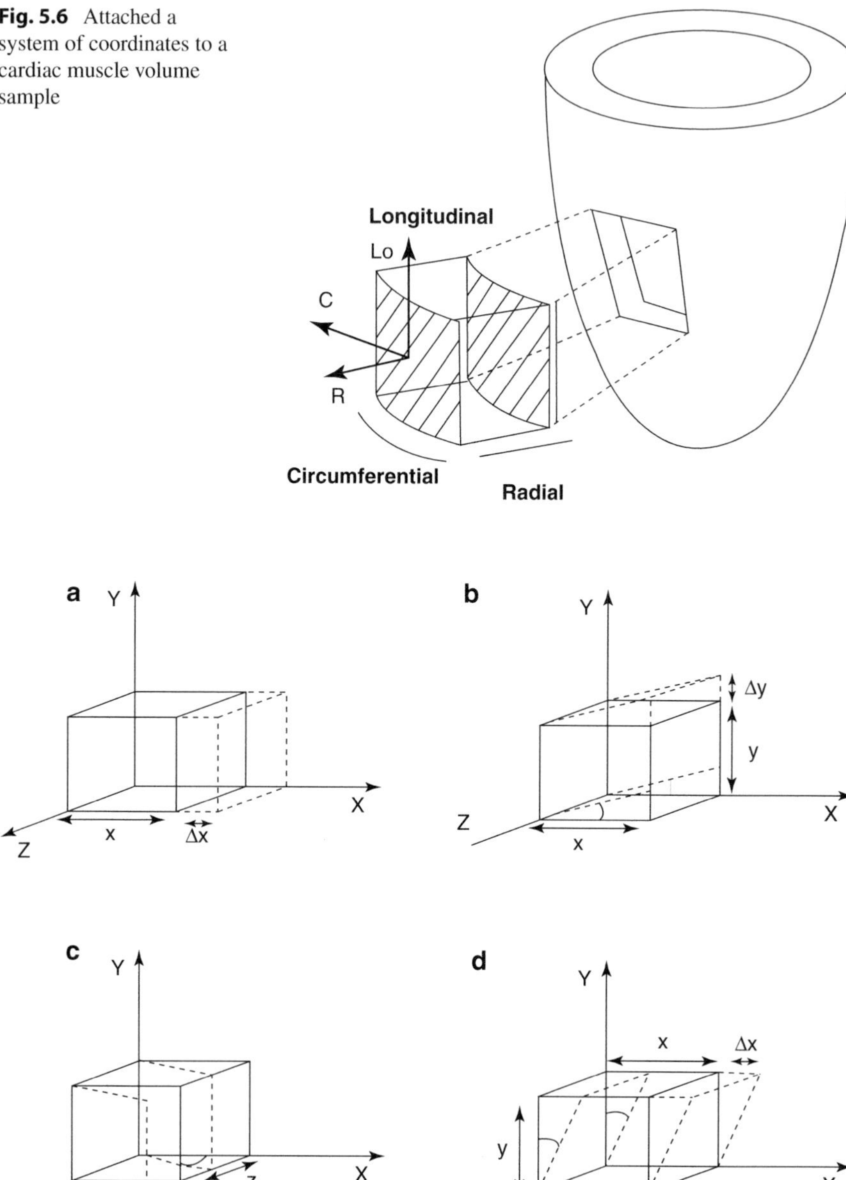

Fig. 5.7 3D changes of a cardiac muscle volume sample respect to the attached system of coordinates of a myocardial segment (**a-d**)

$$\varepsilon_x = \frac{\Delta x}{x}, \varepsilon_{xy} = \frac{\Delta x}{y}, \varepsilon_{xz} = \frac{\Delta x}{z}.$$

$$\varepsilon_{yx} = \frac{\Delta y}{x}, \varepsilon_y = \frac{\Delta y}{y}, \varepsilon_{yz} = \frac{\Delta y}{z}.$$

$$\varepsilon_{zx} = \frac{\Delta z}{x}, \varepsilon_{zy} = \frac{\Delta z}{y}, \varepsilon_z = \frac{\Delta z}{z}.$$

We name these entries as the 3D strain components by ε_{ij}'s for $1 \leq i, j \leq 9$.

In conclusion, the generalized nine variables ε_{ij}'s for $1 \leq i, j \leq 9$ explain the strain components attached to a cardiac muscle volume sample [4, 5].

References

1. Burckhardt CB. Speckle in ultrasound b-mode scans. IEEE Trans Son Ultrason. 1978;25:1–6.
2. D'Hooge J, Heimdal A, Jamal F. Regional strain and strain rate measurements by cardiac ultrasound: principles, implementation and limitations. Eur J Echocardiogr. 2000;1:54–70.
3. Leitman M, Lysyansky P, Sidenko S, Shir V, Peleg E, Binenbaum M. Two dimensional strain—a novel software for real-time quantitative echocardiographic assessment of myocardial function. J Am Soc Echocardiogr. 2004;17:1–9.
4. Mor-Avi V, Lang RM, Badano LP, Belohlavek M, Cardim NM, Derumeaux G. Current and evolving echocardiographic techniques for the quantitative evaluation of cardiac mechanics: ASE/EAE consensus statement on methodology and indications endorsed by the Japanese Society of Echocardiography. Eur J Echocardiogr. 2011;12:167–205.
5. Otto CM. The practice of clinical echocardiography. Philadelphia, PA: Saundes Elsevier; 2011.

A Novel Mathematical Technique to Assess Left Ventricular Myocardial Forces Based on Echocardiography

In this chapter it will be shown that, according to the scheme theory in algebraic geometry, the human heart as an elastic body can be represented as a 3D scheme, on account of algebraic equations of the myocardial fibers as the local parts of the global scheme. It is possible that the fiber movements to be discussed here are identical with the so-called "myocardial fiber transactions"; however, the information available regarding the latter is lacking in precision, and I cannot form a judgment on the matter. It is hoped that some enquirer may succeed shortly in this introduced scheme suggested here, which is so important in connection with the theory of schemes.

I draw a general algorithm of this idea and I intend to give a brief discussion for each part in the mentioned algorithm step by step [1–3].

1. A regional left ventricular wall motion study based on nonlinear dimensionality reduction methods (utilizing NLDR software).
2. A fibered modeling validation of the left ventricle using reconstructed curves at part 1 (global data).
3. Body forces "F" of landmark and contact points on the obtained fibers using of the mathematical elasticity theory [4].

A Regional Left Ventricular Wall Motion Study Based on Non-linear Dimensionality Reduction Method (Utilizing NLDR Software)

Images were acquired from the end of diastole to the end of systole at different phases:

(1) Isovolumic relaxation time, (2) Rapid filling, (3) Diastasis, (4) Atrial contraction, (5) Isovolumic contraction, (6) Ejection time and (7) the end of systole at 4 apical chamber (4C) and short-axis (SX) views within a cardiac cycle by an echocardiography machine. Obtaining informative images at denoted phases will depend

© Springer Nature Switzerland AG 2023

M. Karvandi, S. Ranjbar, *A Review on Recent Echocardiographic Software*, https://doi.org/10.1007/978-3-031-29046-6_6

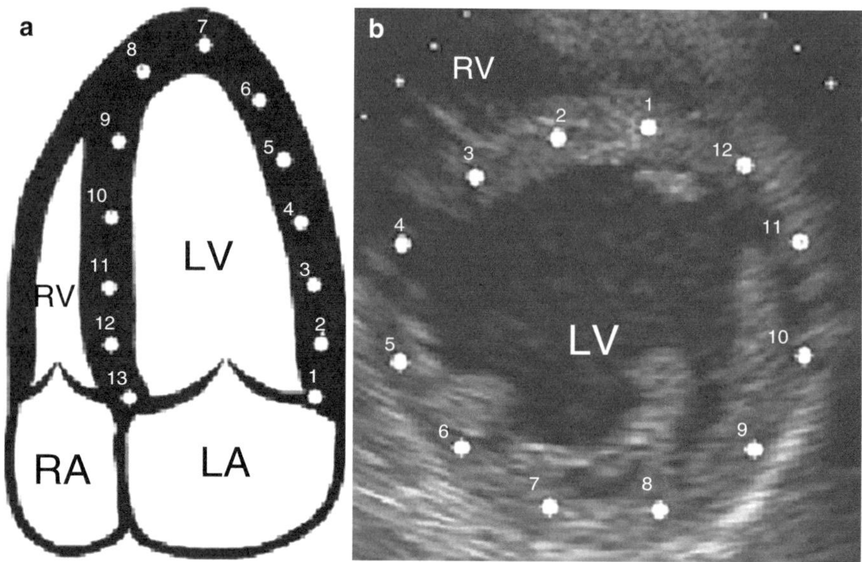

Fig. 6.1 (**a**) LV myocardial region has been divided to 13 material points at a 4C view. (**b**) LV myocardial region has been divided to 12 material points at a SX view

on the operator (for the best clinical examination). Short-axis views are used to evaluate the radial, circumferential, and rotatory data of the left ventricular myocardial sample. The regions of the left ventricular myocardium are divided to 13 points and 12 points at 4C and SX views respectively (Fig. 6.1).

Echocardiography images play the role of observed data. We attach a sequence of images to each LV myocardial divided regions respectively and they are symbolized by $x_{i,r}$'s; in addition, there is a sequence of observed data $x_{1,r}, x_{2,r}, \ldots, x_{N,r}$, where N is the number of obtained images and r is one of the numbered regions. In fact, we have made a chain of displacements, rotations, and pure strain or non-rigid transformations (deformations) of these observed data that is started at $x_{1,r}$ and is terminated at $x_{N,r}$ ($x_{1,r} \to x_{2,r} \ldots \to x_{N-1,r} \to x_{N,r}$). These chains have some conceptual interpretations of the regional LV fiber arrangement/movement [5]. Acquired images $x_{1,r}, x_{2,r}, \ldots, x_{N,r}$ are embedded to a meshed surface-sized number of pixels of an image. Translations, rotations, and pure deformations of $x_{1,r}, x_{2,r}, \ldots, x_{N,r}$ have been implemented and studied at this surface utilizing the mathematical elasticity theory [4]. A distance metric is defined to compute distances between observed points. Medical interpretations of observed points over the time should be checked and stated clearly. Images are not usual images; they are images from the left ventricle. Motion (displacement and velocity), deformation (strain and strain rate), and torsion of each myocardial sample have to be extracted during a cardiac cycle in the mentioned surface. These are used on the creation of a graph G with observed data as its vertices [3, 6].

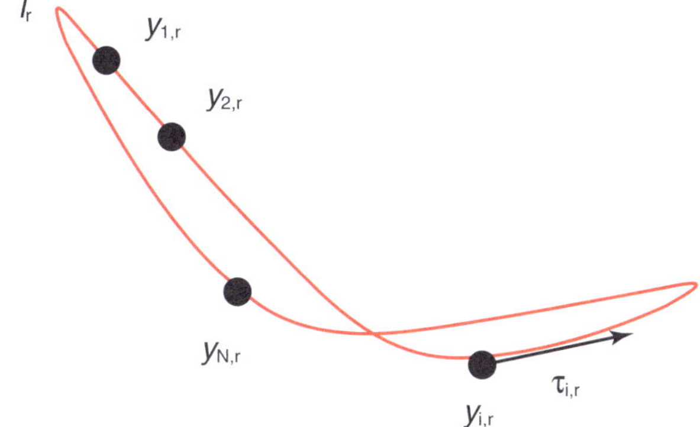

Fig. 6.2 A reconstructed curve l_r on a 3D manifold space and hidden data $y_{i,r}$'s; $f(x_{i,r}) = y_{i,r}$. r is a numbered sample region and $\tau_{i,r}$ is the tangent unit vector in point $y_{i,r}$

The main tool in the NLDR methods is the function "f", which is an isometric/conformal map. Roughly speaking, the embedding "f" is optimized to preserve the local configurations of consecutive data sets [2]. A curve is reconstructed in a 3D manifold space crossing new data $y_{i,r}$'s; $f(x_{i,r}) = y_{i,r}$ (Fig. 6.2).

A Fibered Modeling Validation of the Left Ventricle (Global Data)

Gluing together of reconstructive curves l_r's for each LV divided myocardial regions r, it gives a whole fibered structure of the left ventricle.

Body Force Vectors of Landmark and Contact Points on the Obtained Fibers Using of the Mathematical Elasticity Theory

The physical properties of tissue materials are closely linked to their microstructure. In order to characterize their microstructures as effectively as possible, all fibers in a sample should be individualized. This work mainly aims at developing a new fiber analysis method that segments a 3D mathematical left ventricular modeling based on an echocardiography image into a background and a set of connected components, each representing a single fiber. Properly completed fiber analyses may provide input data for generative or synthetic models, which can, in turn, be used to estimate various characteristics of the material. In this chapter, we introduce an original method based on the skeletonization of the fiber mass, followed by a geometrical analysis of the obtained skeleton. Based on this procedure, several measurements can be computed (e.g. length, translation, orientation, pure deformation, number of contacts, body force). Myocardial point connections together can be realized by a geometrical point of views like a polygon [5] (Fig. 6.3).

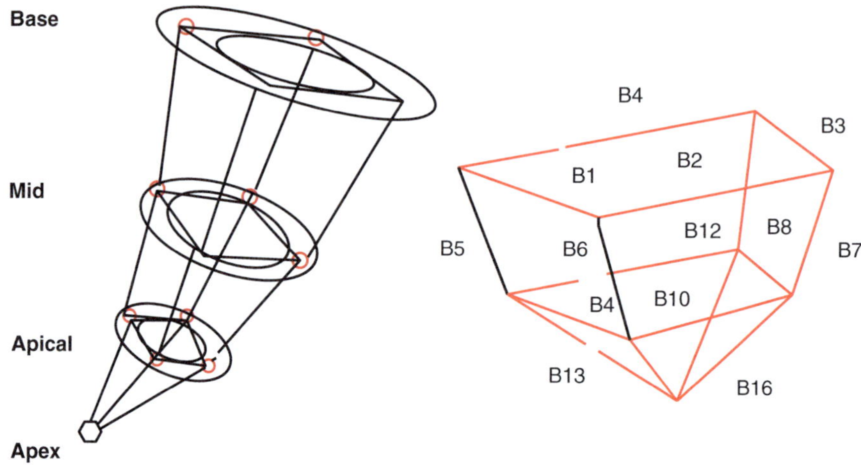

Fig. 6.3 Red points are corresponding geometrical points to material points of LV myocardium. Material points are usually selected as contact points in the myocardium

Figure 6.3 depicts a simplicial complex in which the vertices are geometrical points corresponding to the material points and the edges show how the myocardial points are connected together.

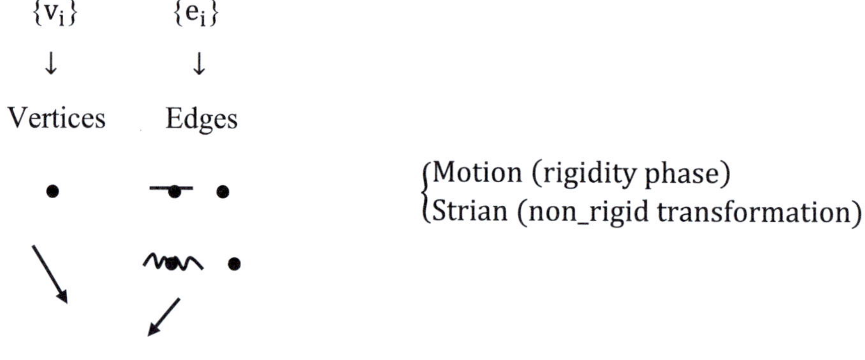

Utilizing reconstructed curves referred to 1, we recover their algebraic equations by the following:

Let p be an arbitrary point on a reconstructed curve $l_{p,r}$ which is presented with respect to the Cartesian coordinates $(x_{1,r}, x_{2,r}, x_{3,r})$ and $\tau_{p,r}$ is the unit tangent vector on it. Body force magnitude F_r at p is reformulated based on pure strain components ($e_{ij,\ell_{p,r}}$) and non-rigid transformations $\tau_{p,r}$ and s (the arc length) respect to the curve l_r by the following formula (Fig. 6.4).

We define $\Gamma_{\ell_{p,r}}$ as the pure tension which acts through the curve $\ell_{p,r}$ on point P by the following formula:

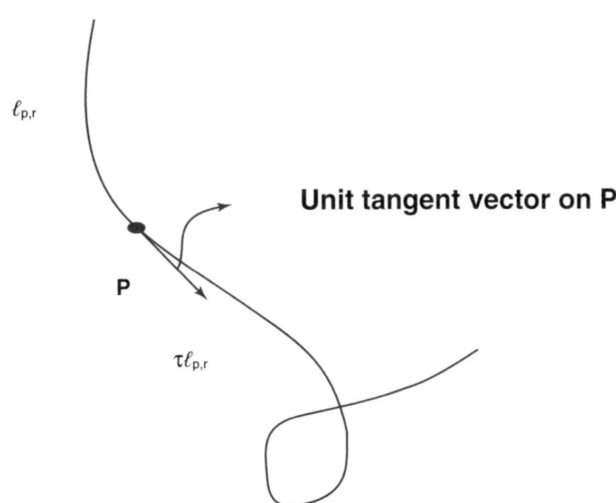

Fig. 6.4 $\ell_{p,\,r}$, the reconstructed curve that acrosses from point P; $\tau_{p,\,r}$, the unit tangent vector on P

$$\Gamma_{\ell_{p,r}} = \sum_{1 \leq i,j \leq 3} e_{ij,\ell_{p,r}} . x_{i,r} . x_{j,r}$$

$e_{ij,\ell_{p,r}}$'s are pure strain components at 3D space they provide a matrix " $\varepsilon = \left[e_{ij\ell_{p,r}} \right]_{3\times 3}$ " which is named the strain matrix.

We can easily formulate the body force over the time at point P on ℓ_P.

$$F_{\ell_{P,r}}(P,t) := \frac{\partial \left(\Gamma_{\ell_{p,r}} \cdot \tau_{\ell_{p,r}} \right)}{\partial S_{\ell_{p,r}}}$$

Now let $\ell^* : = \{\ell$ inside LV myocardium $\mid P \in \ell : = \ell_{p,r}$ for some $r\}$, then the total body force that is resulted in the pure strain and motion is formulated by:

$$F_P(P,t) := \sum_{\ell_{p,r} \in \ell^*} F_{\ell_{p,r}}(P,t)$$

$$\sum_{\ell_{p,r} \in \ell_P} F_{\ell_P}(P,t) = \sum_{\ell_{p,r} \in \ell^*} \frac{\partial \left(\Gamma_{\ell_{p,r}} \cdot \tau_{\ell_{p,r}} \right)}{\partial S_{\ell_{p,r}}} = \sum_{\ell_{p,r} \in \ell^*} \frac{\partial \left(\left(\sum_{1 \leq i,j \leq 3} e_{ij,\ell_{p,r}} x_{i,r} x_{j,r} \right) \cdot \tau_{\ell_{p,r}} \right)}{\partial S_{\ell_{p,r}}}$$

$$F_P(P,t) = \sum_{\ell_{p,r} \in \ell^*} \frac{\partial \left(\left(\sum_{1 \leq i,j \leq 3} e_{ij,p,r} x_i x_j \right) \cdot \tau_{\ell_{p,r}} \right)}{\partial S_{\ell_{p,r}}}$$

Body force at each point P in the LV myocardium are calculated by above formula based on Speckle Tracking software (ST software) to evaluate e_{ij} and velocities and displacements. Let X be the set of all LV myocardial points P. we define the force map φ:

$$\varphi : X \to \mathbb{C}^3$$
$$P \mapsto \left(F_P, \Gamma_P, \tau_P \right)$$

$$\Gamma_P = \sum \Gamma_{\ell_{P,r}} = \sum_{\ell_{P,r}} \sum_{1 \leq i,j \leq 3} e_{ij,\ell_{p,r}} x_{i,r} x_{j,r}$$

$\tau_P \rightarrow$ *Normal vector on the* surface of tangent vector at P. φ is a *smooth* and flat map; $\varphi^{-1}(z)$'s are myocardial fiber transactions for each $z \in C^3$.

$$\varphi^{-1}(z) = \{P \in X \,|\, (F_P, \Gamma_P, \tau_P) = (z_1, z_2, z_3)\}$$

$$\Rightarrow X_{z,p} = \mathrm{Spec}\left(\frac{\mathbb{C}[x,y,z]}{F_P - z_1 \cdot \Gamma_P - z_2 \cdot \tau_P - z_3}\right); \quad z \in \mathbb{C}^3, z_i \in \mathbb{C}.$$

$$X = \coprod_{\substack{z \in \mathbb{C}^3 \\ P \in X}} \mathrm{Spec}\left(\frac{\mathbb{C}[x,y,z]}{F_P - z_1 \cdot \Gamma_P - z_2 \cdot \tau_P - z_3}\right)$$

Finally, the left ventricle "X" is represented by a 3D scheme over complex numbers.

In the next chapter, the fiber movements to be discussed are identical with the so-called "myocardial fiber transactions". It is hoped that some enquirer may succeed shortly in this study here, which is so important in connection with the theory of schemes [7].

References

1. Ming L, Zhen HK. Study of the closing mechanism of natural heart valves. Appl Math Mech. 1986;7:955–64.
2. De Silva V, Tenenbaum JB. Global versus local methods in nonlinear dimensionality reduction. Cambridge, MA: MIT Press NIPS; 2002. p. 705–12.
3. Ranjbar S, Karvandi M. Letter to the editor regarding paper "automatic computation of left ventricular volume changes over a cardiac cycle from echocardiography images by nonlinear dimensionality reduction". J Digit Imaging. 2014;28(2):130–1. https://doi.org/10.1007/s10278-014-9740-x.
4. Sokolnikoff IS. Mathematical theory of elasticity. Malabar, FL: Krieger; 1983.
5. Tenenbaum JB, de Silva V, Langford JC. Global geometric framework for nonlinear dimensionality reduction. Science. 2000;290:2319–23.
6. Saxena A, Gupta A, Mukerjee A. Non-linear dimensionality reduction by locally linear isomaps. In: International Conference on Neural Information Processing (ICONIP), vol. 3316; 2004. p. 1038–43.
7. Ranjbar S, Karvandi M, Ajzachi M. System and method for modeling left ventricle of heart. US Patent Number: 8,414,490. 2013.

Left Ventricular Fiber Arrangements and Orientations

7

Left ventricular torsion from helically oriented myofibers is a key parameter of cardiac performance. Physicians observing heart motion on echocardiograms, during cardiac catheterization, or in the operating room, are impressed by the twisting or rotary motion of the left ventricle during systole. Conceptually, the heart has been treated as a pressure chamber. The rotary or torsional deformation has been poorly understood by basic scientists and has lacked clinical relevance. The aim of this chapter is to discuss this question: Is ventricular twisting related to ventricular fiber arrangement? That is dependent on an assumed model of the left ventricular structure [1–5].

The twisting motion of the heart is believed to be secondary to the arrangement of the muscle fibers. Pettigrew performed careful dissection of the heart of mammals and man, demonstrating seven muscle layers. The three outer layers spiral with an increasing angle from the perpendicular, while the fourth layer is horizontal. The three inner layers spiral in the opposite direction, increasingly toward the vertical. The layers are arranged in opposition so that 1 opposes 7, 2 opposes 6, and 3 opposes 5, with the fourth layer being a connecting layer. Pettigrew postulated that one triplet of muscles contracts during systole and the other stores energy that is utilized in diastole. In his view, the motion of the heart muscle is like that of a torsional pendulum. A reproduction of one of his anatomic dissections can be seen in Fig. 7.1.

Myocardial fiber orientation was examined by Streeter et al. He reported that there was a well-ordered distribution, varying from 60° (from the circumferential axis with positive being toward the base) on the endocardium to approximately −60° on the epicardium of the heart. He found that this fiber angle increased in systole by approximately 7° near the base and 19° near the apex relative to their counterparts in diastole, suggesting a torsion during contraction of die left ventricle [6–8].

More recently, Fernandez-Tran and Hurle confirmed that there are three muscle fiber layers in the heart wall based on orientation. They described superficial (subpicardial), middle and deep (subendocardial), muscle layers which are similar for

© Springer Nature Switzerland AG 2023

M. Karvandi, S. Ranjbar, *A Review on Recent Echocardiographic Software*,

https://doi.org/10.1007/978-3-031-29046-6_7

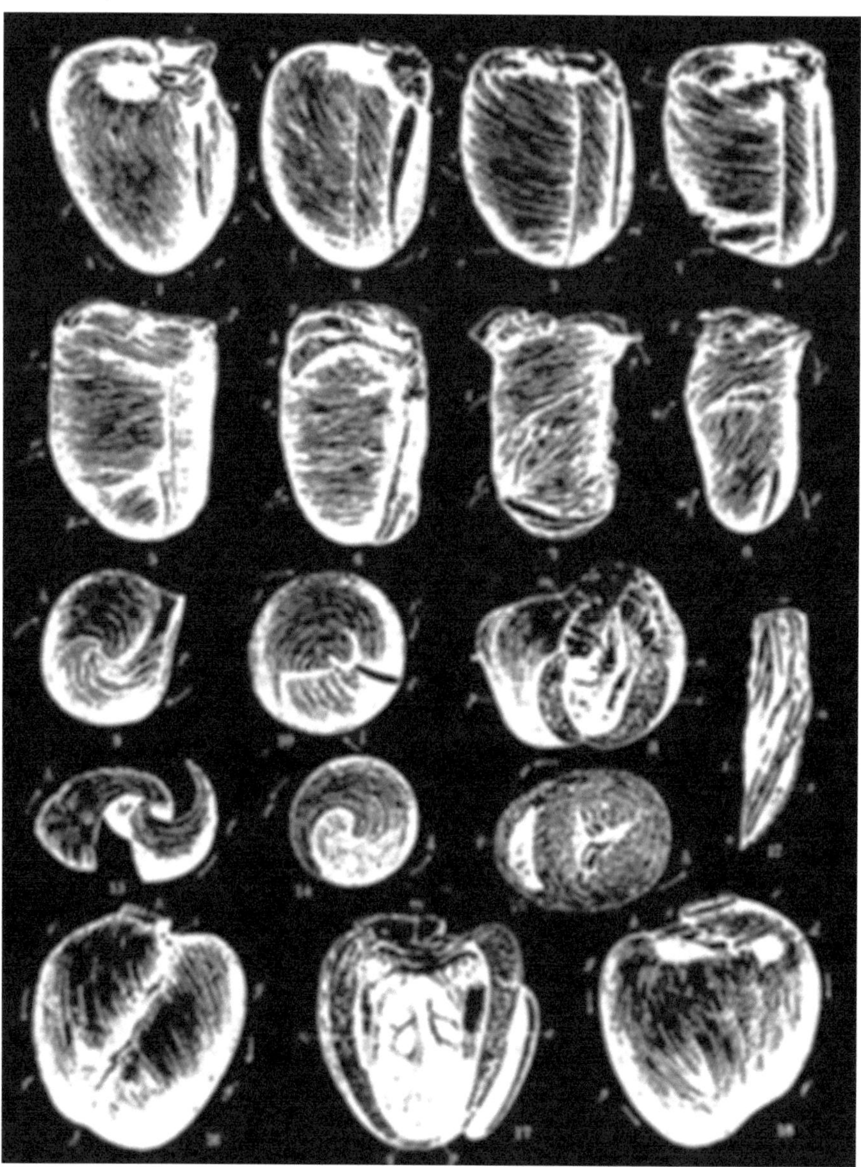

Fig. 7.1 Anatomical dissections

both the left and right ventricle, with the exception of the middle layer which is found only in the left ventricle.

The opposing epicardial and endomyocardial muscle bundles might be important functionally. If potential energy is stored in these muscle fibers, the sudden release might promote filling during early diastole or the period of rapid ventricular filling. The proposed muscle bundles would also give a morphological correlate to the twisting or wringing motion of the left ventricle.

Since the discovery of the helical ventricular myocardial band by Francisco Torrent-Guasp more than 50 years ago [4] and the functional impact of the myocardial band, many scientists have debated the validity of this concept. In any case, Torrent-Guasp's helical spiral concept may find an ideal connection with the spiralization of the outflow tracts and great arteries probably being related to the asymmetric intracardiac flow [7, 8] and/or to the spiral pattern detected at cellular/ molecular level. If the normal rightward spiralization represents the best hydrodynamic pattern, our first aim should be to try to restore the geometric pattern as close as possible to normality, avoiding, as it sometimes happens, the most abstruse surgical choice. In conclusion, the form and function of the heart are inevitably interdependent and we believe this to be true at each phylogenetic and ontogenetic stage. To treat heart diseases as best we can, we must try to understand the structure, function and deep mechanisms of the normal heart from its 'origin'. Once we have acquired as much knowledge as possible in this enormous field, we will then need to do the most difficult thing: mimic nature!

Models of Heart Structure

Eight conceptual models of cardiac structure are summarized in Figs. 7.2 and 7.3. The variety of proposed structures is visually striking; this is due, in large part, to the different levels of cardiac structure described. Model 6 (Fig. 7.2, part 6) groups the myocardium into regional functional units analogous to skeletal muscles. Models 1—5 (Fig. 7.2, parts 1—5) are continuum concepts stressing, to differing degrees, the anisotropic interconnectivity inherent in the cardiac structure. Models 1 and 2 (Fig. 7.2, parts 1 and 2) describe fiber orientation. Model 3 (Fig. 7.2, part 3) describes changes through layers from epicardium to endocardium. Models 4 and 5 (Fig. 7.2, parts 4 and 5) examine the myolaminar structure, with less emphasis on fibre orientation. Some descriptions of Model 7 (Fig. 7.2, part 7), the Helical ventricular myocardial band (HVMB), have emphasized discrete structural bundles while other reports have proposed the band within a continuum framework. Model 8 describes how a dynamic orientation contraction (through the cardiac cycle) of every individual myocardial fiber could be created by adding together the sequential steps of the multiple fragmented sectors of that fiber. This way we attempted to mechanically illustrate the global LV model (Fig. 7.3). We serve to describe explicitly three models of heart that have been provided by the authors.

Fig. 7.2 Mechanisms of heart in different models

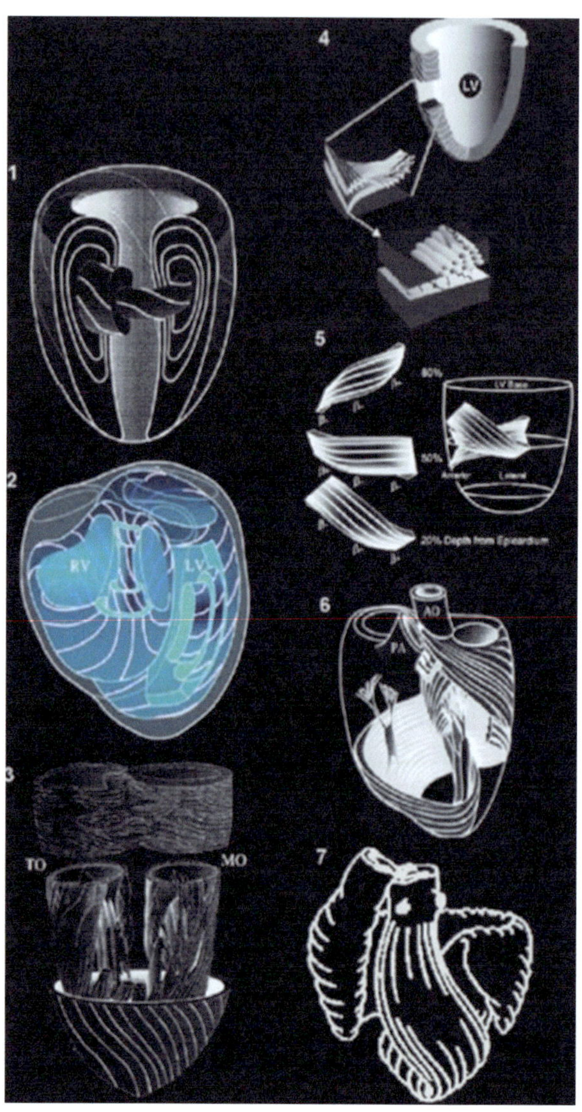

Fig. 7.3 The rout of myofibers in the left ventricle

Doughnut and Pretzel Models: Models 1 and 2 (Fig. 7.2, Parts 1 and 2)

The fiber path models are based on the synthesis of many measurements of angles of fiber orientation through the ventricular wall and differ primarily in their scope (Model 1 is left ventricle (LV) only; Model 2 is LV and right ventricle (RV)). No argument is made for discrete traceable fiber paths—rather a recognizable general fiber trajectory. These models reconstruct the long-accepted helical pattern of fiber orientation, which has recently been confirmed by DT-MRI (Fig. 7.3). DT-MRI fiber tracing algorithms (Zhukov and Barr and Kondratieva et al. applied to canine data; Schmid et al. applied to porcine data; Rohmer et al. applied to human data) and automatic computational constructive visualization methods used to extract meaningful visual information from histologically recorded three-dimensional fiber orientation datasets (recorded from the rabbit heart), produce models remarkably similar to that proposed by Streeter (Fig. 7.2, part 1).

Peskin has carried out an asymptotic analysis of Model 1 and has derived the fiber architecture of the heart from first principles.

These models form a conceptual model of cardiac structure from the one macro- and microscopically observed characteristic of principal fiber direction. As such, the nested doughnuts/pretzel surfaces are abstract concepts no discrete biological equivalents exist. It is possible to create the surfaces by dissection following the observed principal fiber direction, as demonstrated in early studies by Torrent-Guasp; in so doing, however, much information is lost. The dissection and histological methods used do not record data of local branching in directions other than the predominant fiber orientation, so by definition no detail of local tissue organization is modelled. Locally branching fibers are smoothed to a single orientation, which is again smoothed to a global ventricular fiber orientation. With reference to Grant's principles, these models represent the highest-order schema of whole heart fiber orientation in isolation. It might be assumed that the fiber maps produced by algorithmic analysis and computer visualization may be more than a conceptual model, but the considerations above apply equally. The DT-MRI fiber-tracing algorithms track the principal fiber orientation from the primary eigenvector; however, the automatic computational constructive methods were applied to histological fiber orientation datasets which only record principal fiber orientation. Being limited to tracking the principal fiber orientation alone, these methods cannot reconstruct any detail of myolaminar structure, and as such reproduce idealized fiber-tracing dissections. Much evidence points to myolaminae as the central feature of the wall motion mechanism. As such, the doughnut and pretzel models represent geometric abstractions of cardiac structure. Their primary uses may be in: (1) refining more histologically detailed models for the constraint of fiber orientation; and (2) modeling the spread of the cardiac action potential, the conduction of which is significantly influenced by principal fiber direction.

Model 8 (Fig. 7.3)

This model provides a dynamic orientation contraction (through the cardiac cycle) of every individual myocardial fiber could be created by adding together the sequential steps of the multiple fragmented sectors of that fiber. Our study shows that in normal cases myocardial fiber paths initiate from the posterior-basal region of the heart, continues through the LV free wall, reaches the septum, loops around the apex, ascends, and ends at the superior-anterior edge of LV [9, 10].

Philosophical Approach for the Description of a Complex Structure

How can so many models co-exist? Some must be wrong, or is it possible that different models are true representations of the heart when considered from different perspectives?

One 1965 review on cardiac structure stands out as a philosophical foundation upon which all future studies should be based. Robert P. Grant [5], puts forward the following principles (along with an excellent description of cardiac structure):

1. As evidenced from numerous dissections, a given fiber segment has branching connections with other fiber segments in several different directions—as such, the myocardial structure is a syncytium-like arrangement.
2. The cardiac structure problem is therefore a three-dimensional network problem,

3. Structure understood from the study of the network depends on the level on which the structure is approached; whether or not an attempt is made to describe the predominant behavior in the entire ventricle(s) or, at the other extreme, to describe the average branching from a specific location,

4. Many models of heart structure can therefore be proposed depending on the conceptual approach,

5. Statistical study of the branching must be added to geometry for an accurate picture of myocardial architecture,

6. In consideration of such statistical problems, one approach is to construct models for different degrees of generalization of the problem,

7. No single model gives the whole story, but together they provide a schema,

8. The existence of a syncytium-like arrangement does not in itself dictate that no separate 'bundles' are present; however, due to the complex structure it is possible to construct by dissection bizarre arrangements which have no underlying anatomical reality.

9. Any dissection of the myocardium may represent: (a) a valid schema from an infinite set of valid schema; (b) a bizarre and meaningless pathway; or (c) a grouping of fiber paths within the syncytium of such general shared fiber direction that it can be considered a physiological or anatomical bundle,

10. A unique anatomical bundle is not the same as a unique physiological bundle—a connection between separate anatomical structures may produce one physiological structure,

11. A statistical approach is required to demonstrate any independent anatomical entity.

When viewed from this perspective it is unsurprising that many models exist, that these are not all mutually compatible and that argument continues. Further evidence for these principles is presented in a dissection study by Fox and Hutchins, where principle (1) is re-stated and emphasized: 'The only level of the network of cells that can be referred to accurately as a fiber is a single cell. The 'fiber' is often only one cell in length before it splits and branches.' When considering the structural debate it is logical to return to Grant's principles—an approach adopted in this chapter. It should be noted that although Grant suspected regional variations in fiber branching within the left ventricle, he was special whether these local prevalences would statistically warrant consideration as separate bundles. One feature of the myocardial structure problem Grant did not consider was that significant structural differences may exist between individuals of the same species.

Ultimately, studies to date indicate that the left ventricle undergoes a torsional deformation, twisting in a counterclockwise direction during systole and then returning in a clockwise direction during diastole. LV myocardial models enable physicians to diagnose and follow up many cardiac diseases and the torsional, deformation is likely related to the myocardial models. The opposing bands of muscle in the ventricular wall appear to be morphologically correlated with function.

References

1. Arts T, Costa KD, Covell JW, McCulloch AD. Relating myocardial laminar architecture to shear strain and muscle fiber orientation. Am J Physiol Heart Circ Physiol. 2001;280:H2222–9.
2. Spotnitz HM, Spotnitz WD, Cottrell TS, Spiro D, Sonnenblick EH. Cellular basis for volume related wall thickness changes in the rat left ventricle. J Mol Cell Cardiol. 1974;6:317–31.
3. Clayton RH. Computational models of normal and abnormal action potential propagation in cardiac tissue: linking experimental and clinical cardiology. Physiol Meas. 2001;22:R15–34.
4. Ranjbar S, Karvandi M, Ajzachi M. System and method modeling left ventricle of heart. US Patent. Patent number: 8,414,490. 2013.
5. Grant RP. Notes on the muscular architecture of the left ventricle. Circulation. 1965;32:301–8.
6. Fox CC, Hutchins GM. The architecture of the human ventricular myocardium. Johns Hopkins Med J. 1972;130:289–99.
7. Vasudevan V, Wiputra H, Hwai Yap Ch. Torsional motion of the left ventricle does not affect ventricular fluid dynamics of both foetal and adult hearts. J Biomech. 2019;96:109357.
8. Amodeo A, Oliverio M, Versacci P, Marino B. Spiral shapes in heart and shells: when form and function do matter. Eur J Cardiothorac Surg. 2012;41:473–5.
9. Gilbert SH, Benson AP, Li P, Holden AV. Regional localization of left ventricular sheet structure: integration with current models of cardiac fiber, sheet and band structure. Eur J Cardiothorac Surg. 2007;32:231–49.
10. Karvandi M, Ranjbar S, Hassantash SA. Mechanical mitral valve modeling: advancing the field through emerging science. Int J Med Imaging. 2014;2(2):24–8.

8

Left Ventricular Myocardial Force Vectors: Mathematical Work Lab

Force imaging with stimulated MATLAB 7.0.4 (MathWorks, Natick, MA) software provides high spatial resolution measurement of three-dimensional myocardial points tracking (Lagrangian displacement) over the entire cardiac cycle. Echocardiography was performed on 70 healthy volunteers. Data evaluated included: velocity (radial, longitudinal, rotational and vector point), displacement (longitudinal and rotational), strain rate (longitudinal and circumferential) and strain (radial, longitudinal and circumferential) of all 16 LV myocardial segments that were prospectively acquired on Vivid E9 with 4 V probe (GH Healthcare, Horton, N). All data sets comprise a force vector field from end diastole to the end systole, when the myocardium contraction is at its maximum [1–3] (Fig. 8.1a, b).

The data set that covers the entire left ventricle consists of short-axis (SA) images captured from the hearts of 70 volunteers. The vector position of spatially design a muscle volume elements was mapped with a starting time at end diastole and time steps 70 ms apart. These positions were captured and registered relative to the positions of these elements at the end systole. The end systole time was determined through echocardiographic images and for this subject was 380 ms after QRS complex. The imaging parameters are as follows: repetition time = 3.1 ms, mixing time = 250 ms, flip angle = 90°, in encoding strength = 6.25 mm/π, out-of encoding strength = 3.21 mm/π, number of averages = 3, number of phases = 3, in-plane resolution = 1.5 × 1.5, slice thickness = 5 mm, and distance factor = 50%. Using the MATLAB view software, the three-vector field was generated in a matrix format.

Segmentation was then performed by masking all parts of the anatomy except for the myocardium. For this study, we have masked regions of the heart outside the left ventricle. Phase unwrapping was then performed on the segmented images by scanning the myocardium area while searching for sudden changes in the force magnitude. These phase wrappings were later unwrapped by adding or sur-tracking the force correction vector value, which corresponds to the 2π radiant changes in phase. This step was repeated separately for all three directions of force vector; MATLAB was used for the calculations. In Fig. 8.2 we can present arbitrary numbers SA slices of the heart at end systole, along with arrows that show the force vector during the

M. Karvandi, S. Ranjbar, *A Review on Recent Echocardiographic Software*,
https://doi.org/10.1007/978-3-031-29046-6_8

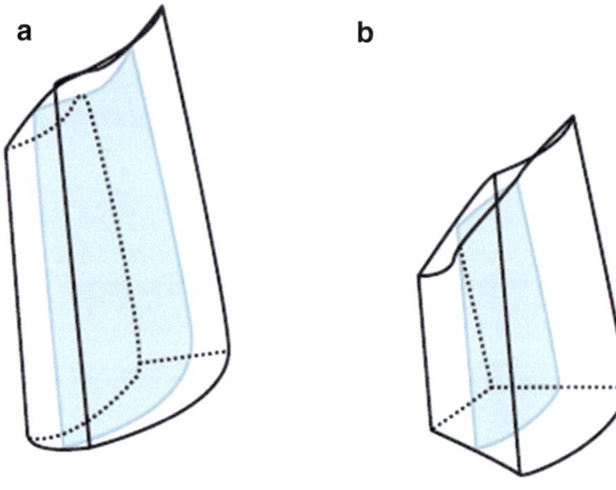

Fig. 8.1 (**a**) Demonstrates a myocardial segment at the end of diastole. (**b**) Demonstrates the same myocardial segment at the end of systole

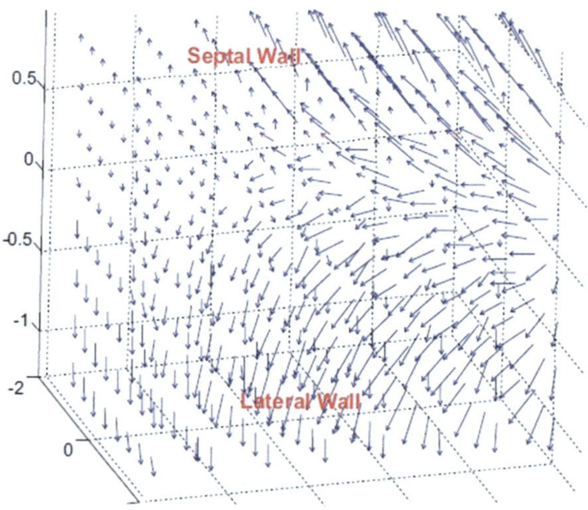

Fig. 8.2 The force vector field on a parasternal long-axis view in MATLAB software

contraction that spans from end diastole to end systole. By having data points across the left ventricle wall, depending upon the wall thickness, we are able to calculate the transmural changes of the thickening and shortening index across the wall and orientation better illustrated in Fig. 8.2. It should be noted that the direction of these vectors repress resulted from the contraction of many myofibers. Therefore, vector directions are not necessarily aligned in fiber directions at each myocardium point. The reduction of the left ventricle volume in systole, and therefore its pumping function, is mostly caused by wall thickening, which is the effect of tangential shortening of the myocardium. Therefore, these two dependent quantities can be used as the quantitative characteristics for the local contribution of the left ventricle myocardium to global heart function [4–7].

The spatial distribution of regions that contribute the most to cardiac function acts macrostructure for the myocardium. The mere existence of such a distinct structure and the knowledge of its normal morphology will facilitate a more effective modeling of the left ventricle function (Figs. 8.3, 8.4, 8.5, 8.6 and 8.7).

Assembled from data shown in Figs. 8.3, 8.4, 8.5, 8.6 and 8.7, we can illustrate a segment of the left ventricle with tracking results from all areas through the heart wall from endocardium to epicardium (Fig. 8.8).

We have suggested a scalar quantity to measure the tangential shortening in a manner that is not sensitive to the deformations in the tangent plane. This quantity, known as the shortening index (SI), measures the change of the in-plane area based on the stretch tensor U, in the radial, longitudinal, and circumferential (RLC) coordinate system. Given a consideration of the incompressibility of the

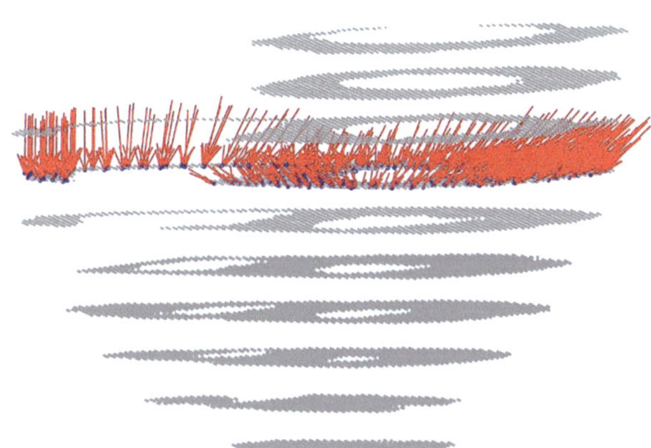

Fig. 8.3 10 short-axis slices are shown at the end systole. For 1 slice, the force field from end diastole to end systole was shown by red arrows

Fig. 8.4 Color-coded force field illustration of all 10 slices

Fig. 8.5 Force field on a long-axis view was depicted here

Fig. 8.6   Side view of the displacement field for 3 slices during the contraction spanning from end diastole to end systole

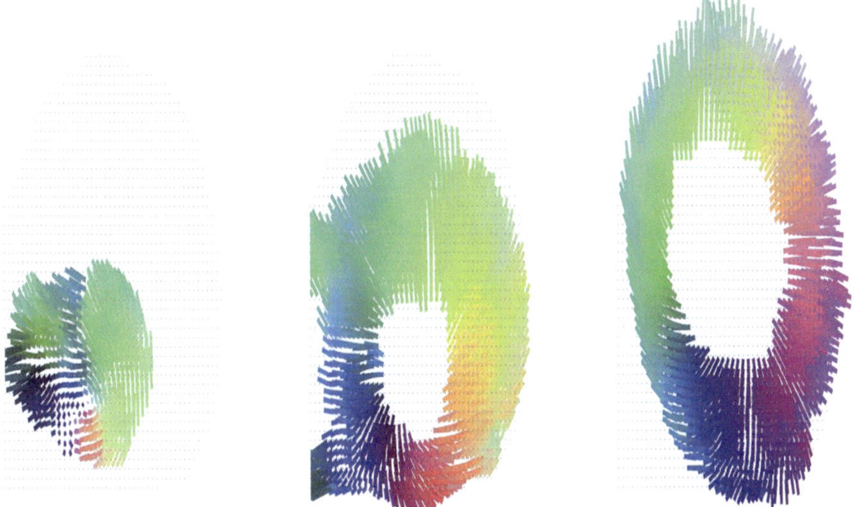

Fig. 8.7 Corresponding top view for the same slices. Color coding was performed to show the direction of vectors

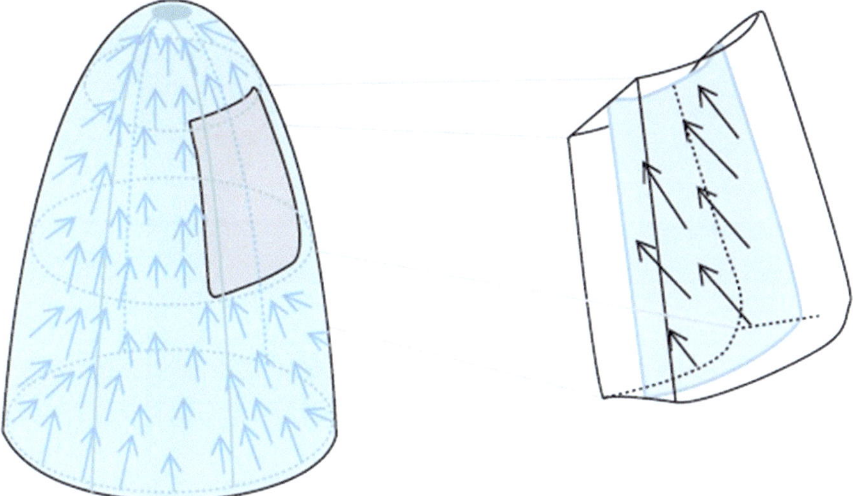

Fig. 8.8 Illustration of the averaged myocardial mesh model of the left ventricle. The arrows indicate the estimated force vector from one frame to the next

myocardium, the local thickening of the wall (T) is also calculated by elements of U. Therefore, T and SI can be measured at any point inside the heart muscle using stretch tensor in MATLAB software. These two dependent quantities have been used in this study as surrogates for the regional contraction level. With the use of the high-resolution force field of the myocardium, maps of T and SI with high spatial resolution were calculated based on the maximum deformation of the left ventricle, i.e., from the end diastole to the end systole. The macrostructure of the left ventricle was subsequently sought by the calculation and comparison of the myofibers of these quantities for their relatively higher absolute values. The myofibers were calculated for SI values above a certain threshold. Starting from low values near zero, we gradually increased the magnitude of the threshold and removed myocardium regions that were left out of myofibers corresponding to the threshold. According to Moore et al. the maximum mean SI magnitude in the left ventricle is $-0.25\ 0.05$ in healthy hearts [5–13]. Figures 8.9, 8.10, 8.11 and 8.12 show myofiber transactions of left ventricle wall thickening and tangential shortening, corresponding to, assembled from data shown in (Figs. 8.3, 8.4, 8.5 and 8.7) that myofibres move on helical bands which begin from the Septal, loop around the Apex and then go to the Anterior.

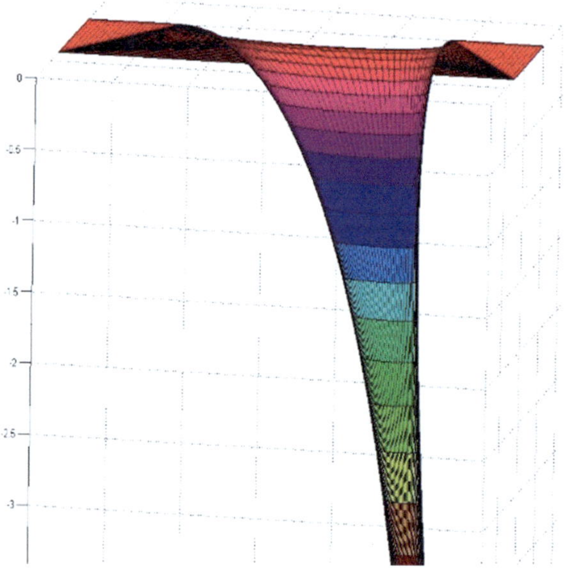

Fig. 8.9 The rout of a myocardial fiber in LV on a long-axis view was depicted here

Fig. 8.10 Myofibers of
left ventricular wall
thickening and tangential
shortening near the
myocardial septum in
MATLAB software

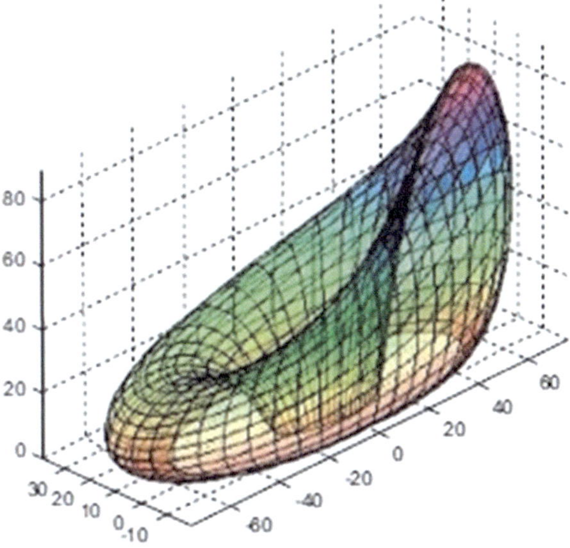

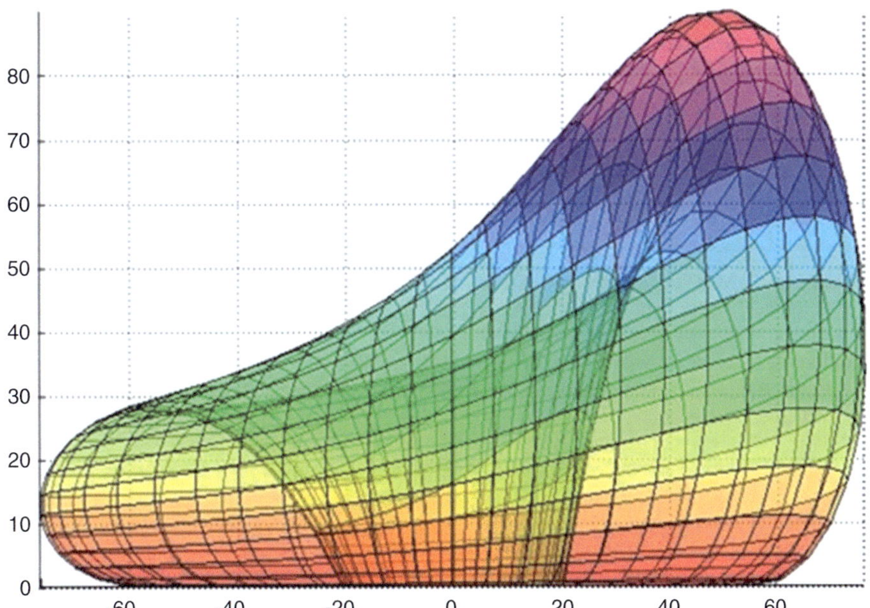

Fig. 8.11 The same myofiber transactions near the myocardial lateral wall in MATLAB software. Finally, we are able to define the whole LV myocardial model mathematically, by MATLAB software in normal subjects [14–17] (Fig. 8.11)

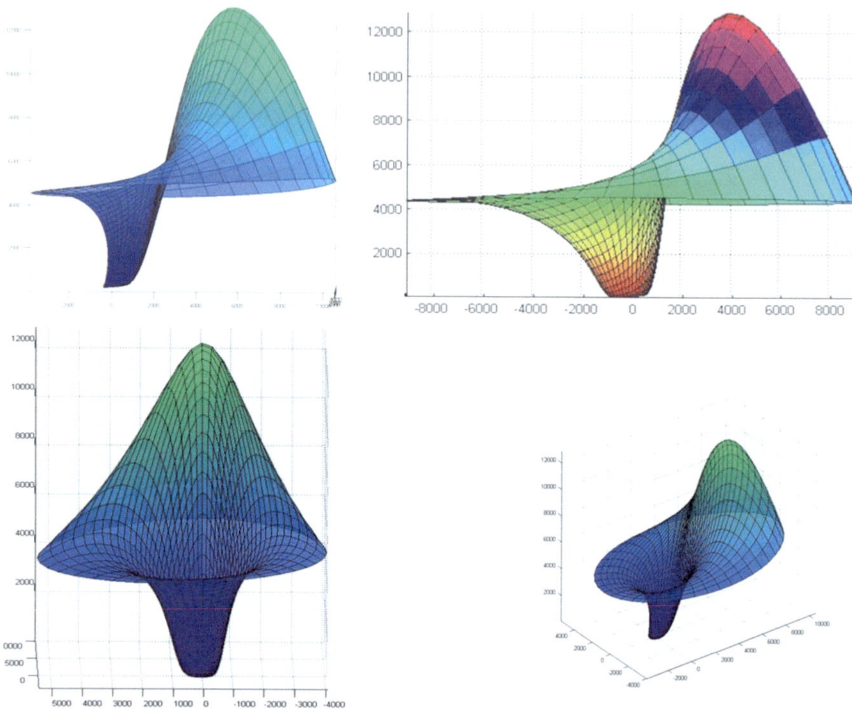

Fig. 8.12 Mathematical modeling of LV related to myocardial fiber paths

References

1. Nesser HJ, Mor-Avi V, Gorissen W, Weinert L, Steringer-Mascherbauer R, Niel J. Quantification of left ventricular volumes using three-dimensional echocardiographic speckle tracking: comparison with MRI. Eur Heart J. 2009;30:65–73.
2. Tanaka H, Hara H, Saba S, Gorcsan J III. Usefulness of three-dimensional speckle tracking strain to quantify dyssynchrony and the site of latest mechanical activation. Am J Cardiol. 2010;105:35–42.
3. D'Hooge J, Heimdal A, Jamal F. Regional strain and strain rate measurements by cardiac ultrasound: principles, implementation and limitations. Eur J Echocardiogr. 2000;1:54–70.
4. Urheim S, Edvardsen T, Torp H, Angelsen B, Smiseth OA. Myocardial strain by Doppler echocardiography. Validation of a new method to quantify regional myocardial function. Circulation. 2000;102:58–64.
5. Leitman M, Lysyansky P, Sidenko S, Shir V, Peleg E, Binenbaum M. Two dimensional strain—a novel software for real-time quantitative echocardiographic assessment of myocardial function. J Am Soc Echocardiogr. 2004;17:1–9.
6. Pirat B, Khoury DS, Hartley CJ, Tiller L, Rao L, Schulz DG. A novel feature tracking echocardiographic method for the quantitation of regional myocardial function: validation in an animal model of ischemia-reperfusion. J Am Coll Cardiol. 2008;51:1–9.

7. Mor-Avi V, Lang RM, Badano LP, Belohlavek M, Cardim NM, Derumeaux G. Current and evolving echocardiographic techniques for the quantitative evaluation of cardiac mechanics: ASE/EAE consensus statement on methodology and indications endorsed by the Japanese Society of Echocardiography. Eur J Echocardiogr. 2011;12:167–205.

8. Helle-Valle T, Crosby J, Edvardsen T, Lyseggen E, Amundsen BH, Smith HJ. New non-invasive method for assessment of left ventricular rotation: speckle tracking echocardiography. Circulation. 2005;112:49–56.

9. Opdahl A, Helle-Valle T, Remme EW, Vartdal T, Pettersen E, Lunde K. Apical rotation by speckle tracking echocardiography: a simplified bedside index of left ventricular twist. J Am Soc Echocardiogr. 2008;21:1–8.

10. Kim DH, Kim HK, Kim MK, Chang SA, Kim YJ, Kim MA. Velocity vector imaging in the measurement of left ventricular twist mechanics: head-to-head one way comparison between speckle tracking echocardiography and velocity vector imaging. J Am Soc Echocardiogr. 2009;22(44):52.

11. Kim HK, Sohn DW, Lee SE, Choi SY, Park JS, Kim YJ. Assessment of left ventricular rotation and torsion with two-dimensional speckle tracking echocardiography. J Am Soc Echocardiogr. 2007;20:45–53.

12. Zhou Z, Ashraf M, Hu D, Dai X, Xu Y, Kenny B. Three-dimensional speckle tracking imaging for left ventricular rotation measurement. An in vitro validation study. J Ultrasound Med. 2010;29:3–9.

13. LeGrice IJ, Takayama Y, Covell JW. Transverse shear along myocardial cleavage planes provides a mechanism for normal systolic wall thickening. Circ Res. 1995;77:182–93.

14. Seo Y, Ishizu T, Enomoto Y, Sugimori H, Yamamoto M, Machino T. Validation of 3- dimensional speckle tracking imaging to quantify regional myocardial deformation. Circ Cardiovasc Imaging. 2009;2:1–9.

15. Moore CC, Lugo-Olivieri CH, McVeigh ER, Zerhouni EA. Three dimensional systolic strain patterns in the normal human left ventricle: characterization with tagged MR imaging. Radiology. 2000;214:453–66.

16. Truesdell CA. A critical summary of developments in nonlinear elasticity. J Ration Mech. 1952;1:125–300.

17. Truesdell CA. A critical summary of developments in nonlinear elasticity. J Ration Mech. 1953;2:593–616.

Lagrangian Mechanic of the Left Ventricle

9

Lagrangian mechanics describes motion in a mechanical system by means of the configuration space. The configuration space of a mechanical system has the structure of differentiable manifold, on which its group of diffeomorphism acts. The basic ideas and theorems of Lagrangian mechanics are invariant under this group, even if formulated in terms of local coordinates.

The Lagrangian point of view allows us to solve completely a series of important mechanical problems, including problems in the theory of small oscillations and in the dynamics of a rigid body.

We reconstructed the left ventricular myocardium as a differential 3D manifold in Chap. 6. Its group of diffeomorphisms are realized by myocardial transformation and deformation over the time that distances, curvatures and energy expansion are remained invariant through the history of the body (Figs. 9.1 and 9.2).

This chapter presents a method for three-dimensional surface flattening, which can be efficiently used in three-dimensional computer-aided garment design. First, a facet model is used to present a complex surface. Then, a spring-mass model based on energy function is used to flatten the 3D mesh surfaces into 2D patterns (Fig. 9.3). The surface elastic deformation energy distribution is depicted by a color graph, which determines a surface cutting line. The method presented here can efficiently solve flattening problems for complex surfaces. The accuracy of a developed surface can easily be controlled locally. Thus, compared to earlier methods, this method provides more flexibility for solving CAD and CAM problems.

Regarding Chap. 2, we discuss several approaches to effectively reduce the computational effort involved in Angle Based Flattening (ABF) and discuss algorithms to effectively solve the parameterization problem. The complexity of the constrained optimization problem raised in the ABF method makes finding a solution in reasonable time a very challenging problem. Several numerical schemes have been proposed to speed up the convergence of the original algorithm by using preconditioning

© Springer Nature Switzerland AG 2023

M. Karvandi, S. Ranjbar, *A Review on Recent Echocardiographic Software*,

https://doi.org/10.1007/978-3-031-29046-6_9

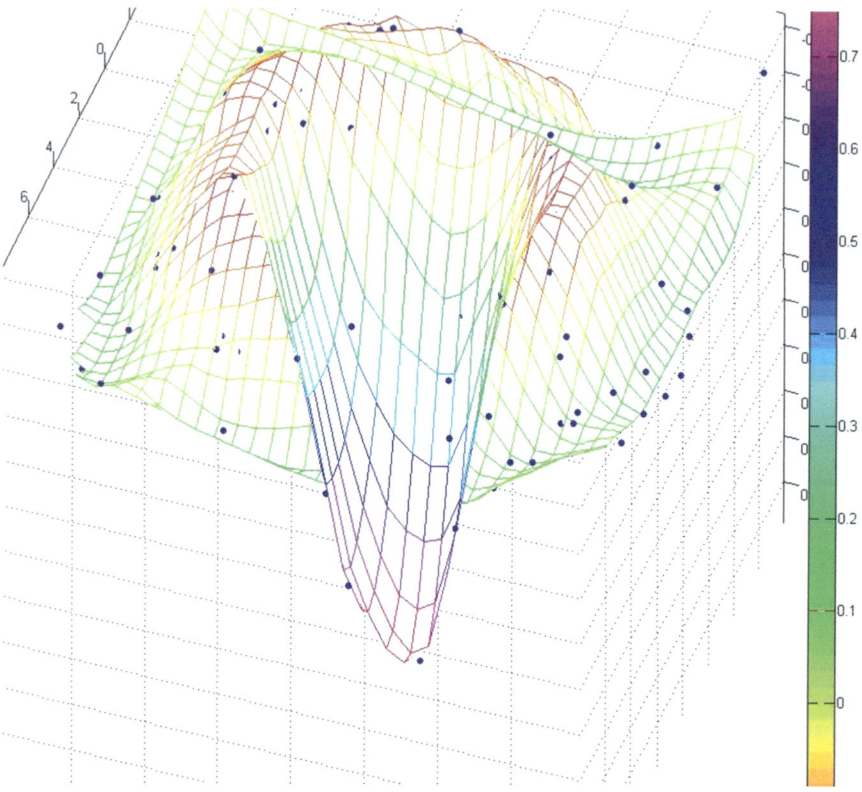

Fig. 9.1 Left ventricle as a differential 3D manifold

and smoothing. We take a completely different approach by identifying the main reasons that hinder convergence within the setting of the constrained problem itself. In fact, the post-processing might be very expensive as it tries to find intersections and then solve the whole nonlinear system as many times as needed. We take advantage of a characterization of convex planar drawings of tri-connected graphs to eliminate boundary intersections in the first place. This way we can steer or even avoid post-processing. Having this characterization in hand, it can be used in association with different objective functions that reflect the criteria we would like to minimize. Such functions can be described as the angular distortion or the **MIPS energy introduced strain matrices** that can be expressed completely in terms of angles [1].

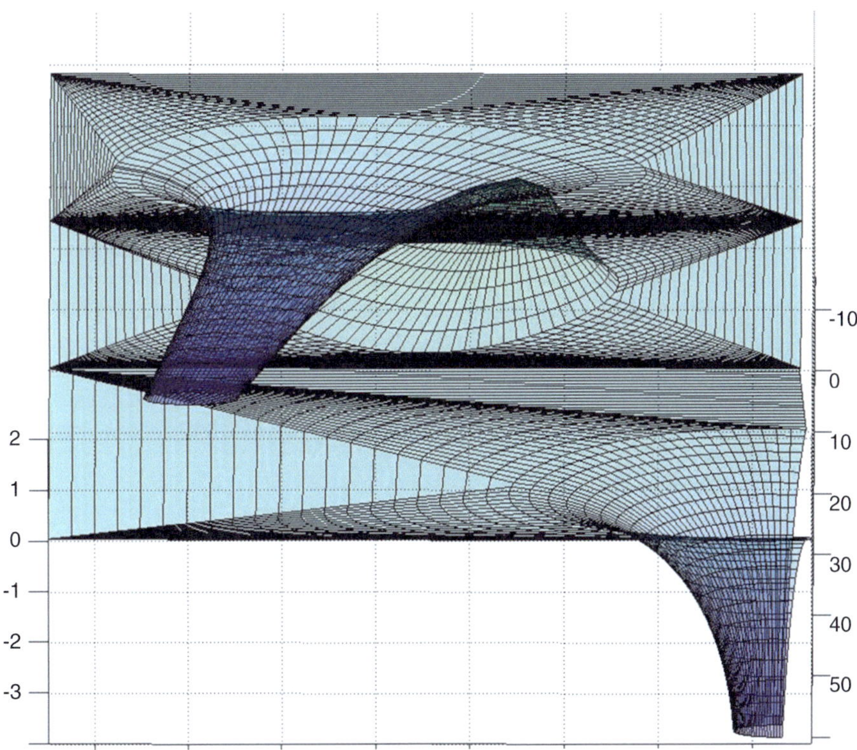

Fig. 9.2 A 3D mesh of the left ventricle of heart as a curvilinear surface

The nonlinear equations in the ABF method lead to a dense sparsity pattern of the Hessian matrix of the system which increases the computational cost. We show how the convergence can be improved alternatively by a simple, yet effective transformation of the problem that relaxes the nonlinear equality constraints. In fact, the Hessian becomes diagonal and its sparsity pattern becomes independent of the valences of the vertices of the input mesh. Since, the system of equations is symmetric we opt for the more appropriate symmetric numerical solvers instead of the non-symmetric ones proposed. We propose a practical approach that achieves fast convergence by finding approximate solutions which yield a low angular distortion.

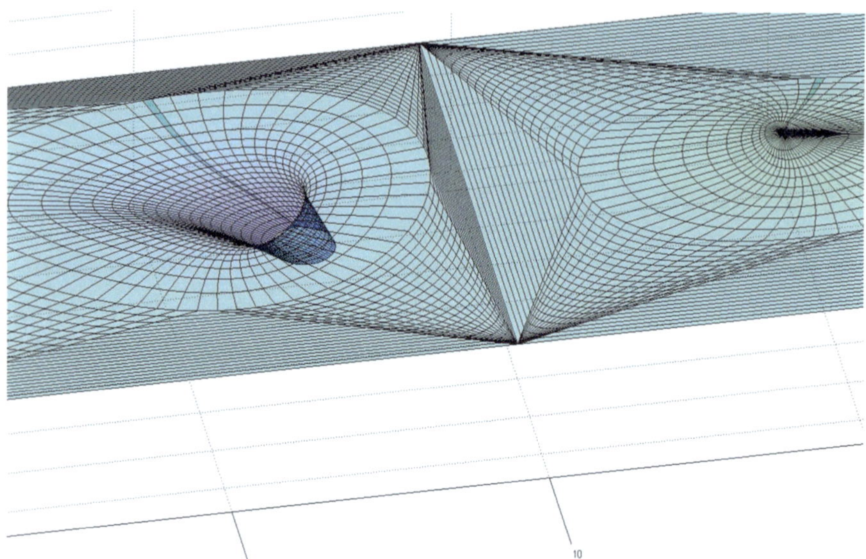

Fig. 9.3 Left ventricular curvilinear surface flattening based on energy model

3D LV Polygon Shape Flattening to Planar Graphs

Chapter 2 addressed the problem of the validity of the planar embedding of the left ventricular polygon by requiring the following consistency condition on the set of positive angles of the planar mesh:

Vertex Consistency

For each internal vertex v, with central angles $\alpha_1, \alpha_2, \ldots, \alpha_d$:

$$\sum_{i=1}^{d} \alpha_i - 2\pi = 0 \tag{9.1}$$

Triangle Consistency

For each triangular face with angles α, β, γ the face consistency:

$$\alpha + \beta + \gamma - \pi = 0 \tag{9.2}$$

Wheel Consistency

For each internal vertex v with left angles $\beta_1, \beta_2, \ldots, \beta_d$ and right angles $\gamma_1, \gamma_2, \ldots, \gamma_d$:

$$\prod_{i=1}^{d} \frac{\sin(\beta_i)}{\sin(\gamma_i)} = 1 \tag{9.3}$$

These conditions guarantee the centered embedding of internal vertices without the overlapping of interior edges. However they do not prevent the overlapping of boundary edges [2, 3]. This issue is a well-studied problem in graph theory.

The minimal constraints for the planarity of the graph impose in addition to (9.1), (9.2), (9.3) the following condition:

Convex External Face Condition

For each external vertex v, with internal angles $\alpha_1, \alpha_2, \ldots, \alpha_d$:

$$\sum_{i=1}^{d} \alpha_i \leq \pi \tag{9.4}$$

Condition (9.4) guarantees the convexity of the boundary and hence prevents boundary overlapping. Note that the inequality (9.4) prevents local and global self-intersection simultaneously. So it does not only prevent adjacent boundary edges from overlapping, but it also guarantees that the boundary loop as a whole does not cross itself. For the local configuration it would in fact be sufficient to require the following weakened condition to hold:

Adjacent Boundary Edges Consistency

$$\sum_{i=1}^{d} \alpha_i \leq 2\pi \tag{9.5}$$

This prevents adjacent boundary triangles from crossing each other. But condition (9.5) is not strong enough to globally enforce a valid mesh with no boundary intersections, as shown in Fig. 9.4.

To get a better understanding and better control of the boundary behavior, we propose multiplying the left-hand side in (9.4) by a positive scalar t, formally

$$\sum_{i=1}^{d} \alpha_i \leq t\pi \tag{9.6}$$

The scalar t can be interpreted as a boundary control coefficient that steers the convexity of the boundary. A lower bound for this factor can be derived using a discrete curvature measure. Consider the angular defect of the flat mesh can be expressed as:

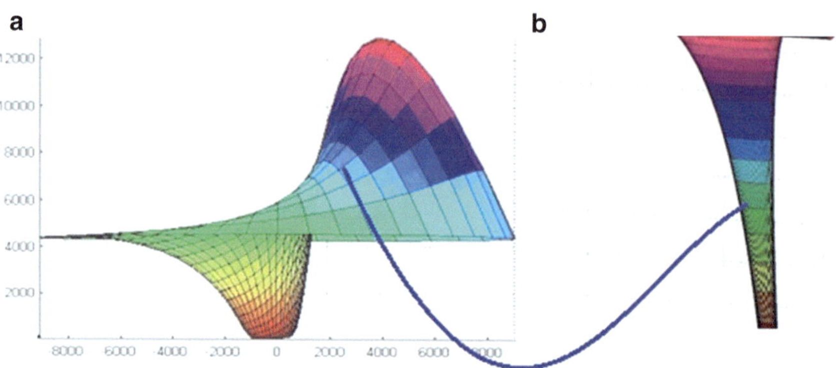

Fig. 9.4 Flattening an α-shaped mode: (**a**) Original mesh. (**b**) The fattened mesh with boundary control coefficient $t = 2$)

$$\sum_{v=1}^{n} \pi - A_v = 2\pi \qquad (9.7)$$

where A_v is the sum of angles at vertex v and n is the number of boundary vertices. By a simple calculation, we establish the lower bound

$$t_0 = 1 - \frac{2}{n}$$

The trivial case is a single triangle, its angles cannot be all smaller than $\frac{\neq}{3}$. We experimented with different values for t, and summarize the following interpretations that can be used as reference for choosing appropriate values for t (Fig. 9.5).

$t > 2$ results in the classic ABF method without preconditioning. No adjacent edge overlapping or boundary self-crossing is taken into consideration.

- $1 < t \leq 2$ prevents adjacent edges from overlapping, but does not necessarily prevent global self-intersections of the boundary loop. We experienced such cases only for "boundary-heavy" (w.r.t. ratio of boundary to inner vertices, e.g. Fig. 9.4) surfaces with non-trivial geometry.
- $t = 1$ globally prevents the boundary loop from self-intersection for any valid input mesh, note that this suffices to induce a convex boundary.
- $t_0 < t < 1$ forces the boundary to become concave.

Figures 9.4 and 9.5 illustrate the behavior of the boundary for different values of t. We can take advantage of these facts in order to avoid an iterative post-processing and thus have better control over the convergence of the constrained optimization problem. In the next sections we show how this problem with the additional inequalities included can be solved efficiently [4].

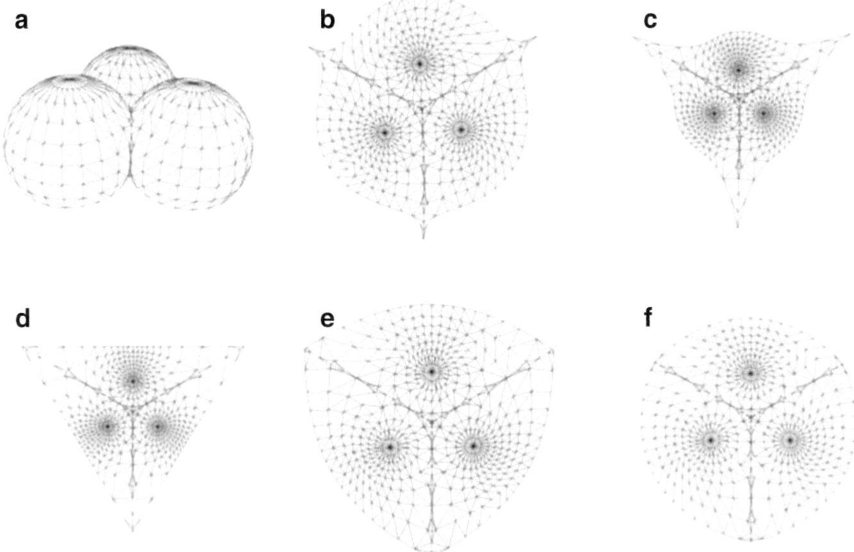

Fig. 9.5 Effect of the boundary control coefficient *t* on the 3-balls model. (**a**) Original mesh. (**b**) Flat mesh for *t* _ 2 (ABF). (**c**) *t* = 1:05. (**d**) *t* = 1 (convex boundary ABF). (**e**) *t* = 0:98. (**f**) *t* = 0:968

Constrained Optimization Problem

A general approach to establish a surface parameterization consists of minimizing an objective function $f(x)$ that quantifies distortion w.r.t. certain quality criteria. The validity of the flat mesh is guaranteed by the angle constraints of **the above section**. A typical choice of such function consists of establishing an angle based objective function. Examples of such functions are the angular distortion

$$f(x) = \sum_{i=1}^{N} w_i \left(x_i - a_i \right)^2$$

With the weights $w_i = \dfrac{1}{a_i^2}$ .The variables a_i represent the optimal angles of the at mesh, which are

$$a_i = \begin{cases} \alpha_i^* \left(2\pi \Big/ \displaystyle\sum_{i=1}^{d} \alpha_i^* \right) & \text{around an interior vertex} \\[2em] \alpha_i^* & \text{around a boundary vertex} \end{cases}$$

Or energy function, that can be written completely in terms of angles

$$f(x) = \sum_{i=1}^{\#\,\text{triangles}} \cot\alpha_i \cot\alpha_i^*$$

We can now formulate the optimization problem as

$$\textit{minimize } f(x)$$
$$\text{subject to } h(x) = 0 \quad g(x) \leq 0 \tag{9.8}$$

Where g and h are multivariate functions of the equality (9.1), (9.2), (9.10) and the inequality constraints (9.6), respectively.

Solving the Optimization Problem

Large constrained optimization systems of the form (9.8) are still open problems in the field of nonlinear optimization. The adequacy of a minimization method depends on the properties of the objective function as well as on the constraints.

In order to solve the optimization problem we use the method of Lagrange multipliers as it guarantees the exact satisfaction of constraints. We handle the inequality constraints by means of the so-called active set approach, a variant of Newton-like methods. It transforms inequalities to equalities which are generally easier to handle. The active set is defined as the set of indices for which the inequality constraint (9.4) is active. Formally

$$A(x,\mu) = \left\{ i \mid g_i \geq \frac{-\mu_i}{c}, i = 1, \ldots, r \right\}$$

where μ_i is the Lagrange multiplier associated with g_i and c is a fixed positive scalar.

The active set approach converts inequality constraints to equality constraints by altering the Lagrange multipliers associated with them. If a constraint does not figure in the active set, its associated multipliers are set to zero. Otherwise it is treated as an equality constraint. The numerical advantage of this method is that as iterates get closer to the solution, the active set becomes more and more stable.

In every Newton iteration the following system is solved

$$\begin{array}{ccc} \nabla^2_{xx}L & J_h^T & J_g^T \\ J_h & 0 & 0 \\ J_g & 0 & 0 \end{array} \times \begin{array}{c} \Delta x \\ \Delta\mu_h \\ \Delta\mu_g \end{array} = - \begin{array}{c} \nabla_x L \\ h \\ g \end{array} \tag{9.9}$$

where the Lagrangian L is given by

$$L = f(x) + \mu_h^T h(x) + \mu_g^T g(x).$$

In the classic ABF algorithm, the computation of the Hessian matrix $\nabla^2_{xx}L$ involves finding the second derivatives of the products involved in condition condition (9.3). The resulting matrix is sparse, but it still contains a considerable number of non-zero elements. This number depends largely on the valences of the input mesh vertices. Instead, we propose to use a modified wheel condition (9.10). Since the angles are strictly positive we can safely rewrite condition (9.3) as

$$\log\left(\prod_{i=1}^{d}\frac{\sin(\beta_i)}{\sin(\gamma_i)}\right) = 0 \text{ or } \sum_{i=1}^{d}\log\left(\sin(\beta_i)\right) - \log\left(\sin(\gamma_i)\right) = 0, \qquad (9.10)$$

The virtue of this modification resides in the fact that it yields a diagonal Hessian matrix.

$$\nabla_{xx}^{2}L = diag\left(f''(x_i) + m_i\left(-\frac{1}{\sin^2(x_i)}\right)\right)$$

where m_i is the linear combination of the Lagrange multipliers involved with x_i in condition (9.10). The amount of computation and effort by the iterative solvers is hence reduced considerably.

Note that the initial guess for the x is the set of the optimal angles $\{a_i | i = 1, 2, ..., n\}$. Consequently, at every Newton iteration the solution stays within the positive domain.

Definition of Lagrangian Function

In this chapter, we use an energy model—planar triangular spring-mass system to obtain the flattening result of a 3D mesh surface of the left ventricle. The procedure of surface flattening is a planar triangular mesh deformation process directed by the Lagrangian function of the spring-mass system.

Generalized Lagrange–Euler Equations Applied to Left Ventricular Myocardial Forces

Lagrangian function of the left ventricle is described in the following way:

$$L\left(P,\dot{P},\varepsilon_P,\dot{\varepsilon}_P,t\right) = T\left(P,\dot{P},\varepsilon_P,\dot{\varepsilon}_{P,t}\right) - U\left(P,\dot{P},\varepsilon_P,\dot{\varepsilon}_P,t\right)$$

$$= \frac{1}{2}V_P(t)\dot{P}^2 - \frac{1}{2}\left(\sum_{\ell_{p,r}\in\ell^*}\left(\sum_{1\le i,j\le 3}e_{ij,\ell_{p,r}}(t)\cdot x_{i,r}\cdot x_{j,r}\right)\right)P^2$$

Such that $V_P(t) = V_{\text{initial}, p} + \delta V_P(t)$ and $\delta V_P(t)$ is the determinant of generalized strain matrix:

$$\begin{matrix} \varepsilon_{P,1} & \varepsilon_{P,2} & \varepsilon_{P,3} \\ \varepsilon_{P,4} & \varepsilon_{P,5} & \varepsilon_{P,6} \\ \varepsilon_{P,7} & \varepsilon_{P,8} & \varepsilon_{P,9} \end{matrix}$$

$P_1, P_2,..., P_9$ are nine position variable toward 9 strain components $\varepsilon_{P,\,1}$, $\varepsilon_{P,\,2}$, $\varepsilon_{P,\,3}, ..., \varepsilon_{P,\,9}$.

Lagrange–Euler Equations

$$\frac{\partial L\left(P_{i,}\,\dot{P}_i,\varepsilon_P,\dot{\varepsilon}_P,t\right)}{\partial P_i} - \frac{\mathrm{d}}{\mathrm{dt}}\left(\frac{\partial L\left(P_i,\dot{P}_i,\varepsilon_{P,i},\dot{\varepsilon}_{P,i,t}\right)}{\dot{P}_i}\right) = F\left(P_i,\dot{P}_i,\varepsilon_{P,i},\dot{\varepsilon}_{P,i},t\right)$$

Since $F\left(P,\dot{P},\varepsilon_{P,i},\dot{\varepsilon}_{P,i},t\right) = \partial\left(\varepsilon_{P,i}\times\dot{P}_i\right)/\partial s_i$ we can rewrite Lagrangian–Euler equations as well:

$$\frac{\partial L\left(P_{i,}\,\dot{P}_i,\varepsilon_P,\dot{\varepsilon}_P,t\right)}{\partial P_i} - \frac{\mathrm{d}}{\mathrm{dt}}\left(\frac{\partial L\left(P_i,\dot{P}_i,\varepsilon_{P,i},\dot{\varepsilon}_{P,i,t}\right)}{\dot{P}_i}\right) = \partial\left(\varepsilon_{P,i}\times\dot{P}_i\right)/\partial s_i$$

We apply kinetic and potential energetic formulas of the left ventricle to the left side of Lagrange–Euler equations

$$\begin{aligned}
&\left[\left[\partial\left[\frac{1}{2}\rho\left(V_{\text{initial},p} + \det\begin{pmatrix}\varepsilon_x & \varepsilon_{xy} & \varepsilon_{xz}\\ \varepsilon_{yx} & \varepsilon_y & \varepsilon_{yz}\\ \varepsilon_{zx} & \varepsilon_{zy} & \varepsilon_z\end{pmatrix}\right)\dot{P}^2 - \frac{1}{2}\left(\sum_{\ell_{p,r}\in\ell^*}\left(\sum_{1\le i,j\le 3}e_{ij,\ell_{p,r}}(t)\cdot x_{i,r}\cdot x_{j,r}\right)\right)P^2\right]/\partial P_i\right]\right.\\
&\left.-\left[d\left[\partial\left[\frac{1}{2}\rho\left(V_{\text{initial},p} + \det\begin{pmatrix}\varepsilon_x & \varepsilon_{xy} & \varepsilon_{xz}\\ \varepsilon_{yx} & \varepsilon_y & \varepsilon_{yz}\\ \varepsilon_{zx} & \varepsilon_{zy} & \varepsilon_z\end{pmatrix}\right)\dot{P}^2 - \frac{1}{2}\left(\sum_{\ell_{p,r}\in\ell^*}\left(\sum_{1\le i,j\le 3}e_{ij,\ell_{p,r}}(t)\cdot x_{i,r}\cdot x_{j,r}\right)\right)P^2\right]/\partial\dot{P}_i\right]/\mathrm{dt}\right]\right) = \partial\left(\varepsilon_{P,i}\times\dot{P}_i\right)/\partial s_i
\end{aligned}$$

Continues solving Lagrangian equations using a speckle tracking method based on echocardiographic initial data, determines the rout of left ventricular fiber bands per a cardiac cycle (Fig. 9.6) [5, 6].

As a conclusion to this chapter, the standard way to model the motion of blood inside the left ventricle would be to treat the left ventricle as an elastic membrane obeying Newton's laws of motion with forces calculated in part from the elasticity of the membrane and, in part, by evaluating the fluid stress tensor on the surface of the membrane. Then the fluid equations would have to be supplemented by the

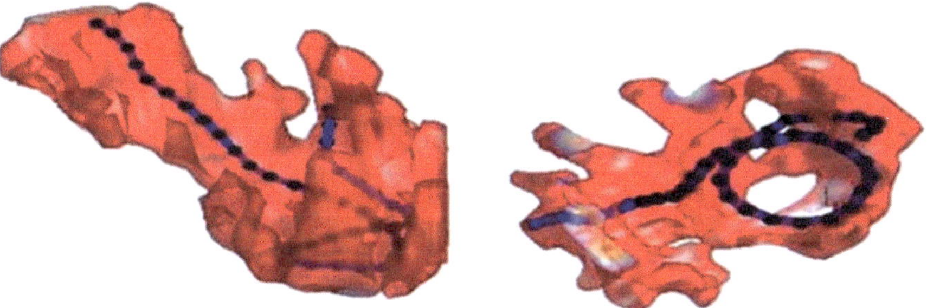

Fig. 9.6 These bands initiate from the posterior-basal region of the heart (denoted by the asterisk), continues through the left ventricle free wall, reaches the septum, loops around the apex, ascends, and ends in the superior-anterior section of the left ventricle

constraint that the velocity of the fluid on either side of the membrane must agree with the instantaneously known velocity of the elastic membrane itself. There is a difficulty with this standard approach to the problem. The challenge is the practical one of evaluating the fluid stress tensor on either side of the boundary. This seems difficult (or at least messy) to do numerically, unless the computational grid is aligned with the boundary. On the other hand, in a moving boundary problem, it is both expensive and complicated to re-compute the grid at every time step in order to achieve alignment. This means that the sum of the elastic force and the fluid force on any part of the boundary has to be zero. Once we know this, it becomes unnecessary to evaluate the fluid stress tensor at the boundary at all! We can find the force of any part of the boundary on the fluid by evaluating the elastic force on that part of the boundary. (Note the use of Newton's third law: the force of boundary on fluid is minus the force of fluid on boundary.) All we need is a method for transferring the elastic force from the boundary to the fluid. On a Cartesian grid, this may be done by spreading each element of the boundary force out over nearby grid points. The particular way that this is done in the boundary method involves a carefully constructed approximation to the Dirac delta function. This force-spreading operation defines a field of force on the Cartesian lattice that is used for the fluid computation. Then the fluid velocity is updated under the influence of that force field. The Navier–Stokes solver that updates the fluid velocity does not know about any consideration of the heart left ventricle geometry; it just works with a force field that happens to be zero everywhere except in the immediate of the vorticity region. This approach can be used for instructional purposes and diagnosis of heart ailments [7, 8].

In conclusion, in this chapter, the relation of the form and function of the left ventricle (myocardium) has been studied through investigating the spatial distribution of its regional function. In this approach, we were able to define the whole LV myocardial model mathematically, by MATLAB software in normal subjects. This will enable physicians to diagnose and follow up many cardiac diseases when this procedure is interfaced within echocardiographic machines.

References

1. Badano LP, Lang RM, Zamorano JL. Textbook of real-time three-dimensional echocardiography. London: Springer; 2011.
2. Helm P, Beg MF, Miller MI, Winslow RL. Measuring and mapping cardiac fiber and laminar architecture using diffusion tensor MR imaging. Ann N Y Acad Sci. 2005;10:296–307.
3. Le Grice I, Hunter P, Young A, Small B. The architecture of the heart: a data-based model. Philos Trans R Soc Lond A Math Phys Sci. 2001;35:1217–32.
4. Ashikaga H, Criscione JC, Omens JH, Covell JW, Ingels NB Jr. Transmural left ventricular mechanics underlying torsional recoil during relaxation. Am J Physiol Heart Circ Physiol. 2004;28:H640–7.
5. Beyar R, Yin FCP, Hausknecht M, Weisfeldt ML, Kass DA. Dependence of left ventricular twist-radial shortening relations on cardiac cycle phase. Am J Physiol Heart Circ Physiol. 1989;257:H1119–26.
6. Ranjbar S, Karvandi M, Ajzachi M. System and method modeling left ventricle of heart. US Patent 8,414,490. 2013.
7. Chorin AJ. Numerical solution of the Navier-stokes equations. Math Comput. 1968;22:745–62.
8. Harlow FH, Welch JE. Numerical calculation of time-dependent viscous incompresible flow of fluid with free surface. Phys Fluids. 1965;8:82–9.

10

Dynamic Features Creating (Which Cause) the Blood Direction Inside the Left Ventricle

The mathematical boundary method is a new technique for determining the velocity and the direction of fluid streams through an analysis of the change in the position of small particles that drift with the fluid. The recent development of echocardiographic technology now makes it possible to apply this approach to echocardiographic imaging.

The growing knowledge about the structure and function of the ventricle was of particular significance to us in the context of ventricular vortex development in the normal subject. A vortex is a mass of fluid with a whirling or circular motion, and thereby containing kinetic energy. It is supposed to increase cardiac efficiency by maintaining the momentum of the inflowing blood in diastole and, thereby, facilitating systolic ejection of blood into the left ventricular outflow tract. The demonstration of diastolic vortex formation in normal human hearts and their distortion by valve surgery led to our study objective of to what extent blood flow patterns in congenitally abnormal, functionally univentricular hearts of patients are different from normal ones. Left ventricular (LV) fluid dynamics represent an elegant flow phenomenon, starting with the trans-mitral jet reaching a maximal velocity of about 100 cm/s in traveling a distance as short as only a few centimeters. This volume of blood must then turn around to exit the left ventricle via the aortic valve within a fraction of a second while maintaining its initial high velocity. For this to happen properly, nature has chosen a vortex flow (a swirling motion of fluid), which is developed during diastole, to redirect the trans-mitral flow toward the LV outflow tract. This track of motion guarantees efficient transfer of the fluid momentum and minimizes the ventricular energy dissipation Despite improvements in intra-ventricular flow visualization by means of advanced noninvasive imaging modalities, experimental and numerical models, several aspects of the corresponding fluid dynamics phenomena are still unclear. The key reason for the lack of quantitative information concerning cardiac flow is the lack of technology that can reliably map the spatial and temporal details of the flow while being feasible in routine clinical use. Color Doppler imaging, as a tool to obtain intra-cardiac flow information,

© Springer Nature Switzerland AG 2023

M. Karvandi, S. Ranjbar, *A Review on Recent Echocardiographic Software*,

https://doi.org/10.1007/978-3-031-29046-6_10

considers only a single component of blood velocity along the transducer scan line and does not provide quantitative information with respect to vortex formation, recirculation areas, and flow stagnation zones, Because of these limitations, color Doppler imaging may have no adequate information for either the reconstruction of velocity vector fields or the actual direction of streamlines. Several attempts have been made to incorporate quantitative fluid dynamics into echocardiography using particle-tracking algorithms. This study aimed at assessing vortex flow patterns in normal subjects circulation in comparison with the other technology [1–3].

A Novel Mathematical Modeling of the Left Ventricular Myocardium

Regarding Chaps. 7–9, Ranjbar et al. (2013) developed the first novel left ventricular myocardial model mathematically based on echocardiography, by MATLAB software and LSDYNA software in normal subjects, in which dynamic orientation contraction (through the cardiac cycle) of every individual myocardial fiber could be created together the sequential steps of the multiple fragmented sectors of that fiber. The left ventricular myocardial modeling of the heart shows that, in normal cases, myocardial fibers initiate from the posterior from the heart, continues through the left ventricular free wall, reaches the septum, loops around the apex, ascends, and ends at the superior-anterior edge of the left ventricle (Fig. 10.1).

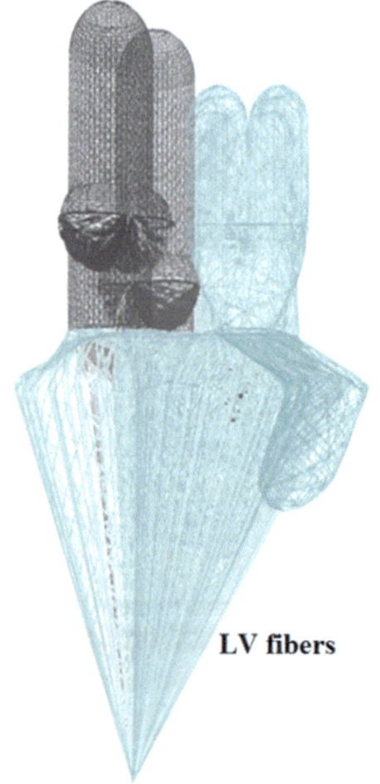

Fig. 10.1 A mathematical fibered modeling of the left ventricle of heart

A Novel Mathematical Technique to Assess of the Mitral Valve Dynamics Based on Echocardiography

The left side of the heart accepts oxygenated blood at low pressure from the lungs into the left atrium. The blood then moves to the left ventricle, which pumps it forward to the aorta to circulate the body. The heart's valves maintain the unidirectional flow of the blood through the heart, that is, the mitral valve prevents the regurgitation of blood from the left ventricle into the left atrium during ventricular systole and the aortic valve prevents blood flowing back from the aorta into the left ventricle during diastole. The principal fluid phenomena involved in the left ventricular diastolic flow are related to the presence of symmetric structures that develop with the strong jet that enters through the mitral valve. It is conjectured that the fibered structure has important effects on the function of the ventricle. The human mitral valve is a complex anatomical structure consisting of two valve leaflets, an annulus, chordate tendineae, and two papillary muscles, which are finger-like projections embedded into the underlying left ventricular myocardium. The mitral annulus is a saddle-shaped fibrous ring which seamlessly transitions into the two leaflets (Fig. 10.2). The leaflets extend into the left ventricle, where they are tethered to the papillary muscles via an intricate arrangement of chordae tendineae. The chordae tendineae consist of a complex web of chords that attach all over the leaflets of the valve. They prevent the prolapse of the valve leaflets at systole, and additionally assist in maintaining the geometry and functionality of the ventricle. The papillary muscles play an important role and are believed to lengthen during isovolumetric contraction and shorten during ejection as well as during isovolumetric relaxation to maintain the chordae at the same deformation level, during opening and closure.

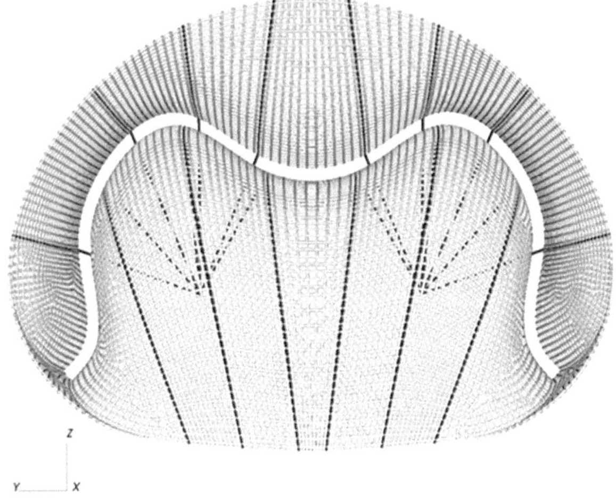

Fig. 10.2 The presentation of the non-planar, annular saddle-shaped of the mitral valve

The mitral valve structure and function is one of the parts in providing/completing for the velocity vector field attached to mitral valve leaflets. Hence, the geometrical mitral valve parameters, including: (1) Anterior tending angle, (2) Anterior leaflet tending angle, (3) Posterior leaflet tending angle, (4) Anterior bending angle, (5) Length of leaflet coaptation, (6) Tending height, (7) Tending area, (8) The length of leaflets, (9) The circumference of the annulus of the mitral valve, (10) Commissure distance, (11) Systolic anterior-posterior distance. (12) Diastolic anterior-posterior distance and (13) Annular height have not been considered at the recent studies in mitral valve leaflets modeling. These measurable echocardiographic data as the elementary data, are at least needed and the main aim of this chapter to reconstruct a mitral valve model to providing a boundary condition of the fluid dynamic of the mitral valve leaflets (Fig. 10.3).

Echocardiography imaging of the mitral valve motion during the cardiac cycle of planes transthoracic echocardiogram (TTE) in the short axis, long axis and four chamber views, and multi-planes transesophageal echocardiogram (TEE) in the lower esophageal views were prospectively acquired for 200 patients of 65 time frames from diastole to systole (early diastole, mid diastole, atrial systole, end of diastole, end of systole). In each plane and for each frame during diastole to systole, 13 mentioned geometrical parameters of mitral valve annulus and leaflets were automatically measured by utilizing TomTec software (Philips Medical, Andover, MA) (Fig. 10.4) [4–10].

When those 13 parameters are assessed then vectors/lines that attach bases, middles and tips of AML and PML together, were manually identified using MATLAB software in the aquaied 65 time frames from diastole to systole [10–14] (Fig. 10.5).

3D AML and PML positions were automatically computed for each frame and used as input to displacement vectors modeling of the mitral valve leaflets (Fig. 10.6).

By solving mathematical equations of inelastic properties of leaflets, each plane was replaced with 60 vectors of displacement and ultimately the mitral valve leaflets were realized by 1125 vectors (Fig. 10.7) of displacements which also show translations, rotations and pure strains of bases, middles and tips of AML and PML simultaneously per cardiac cycle [15–18].

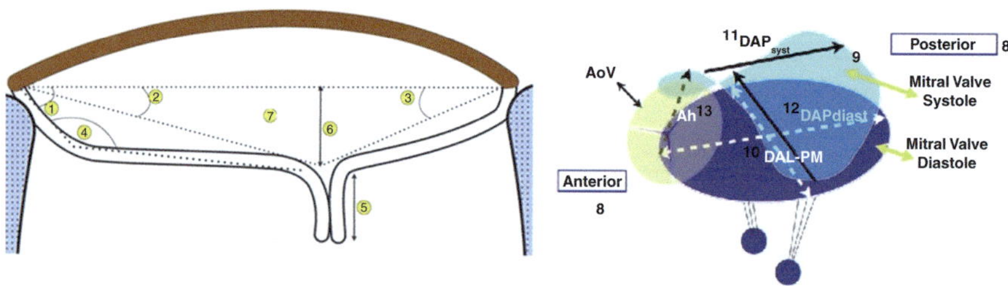

Fig. 10.3 The 13 parameters used to define the geometry of mitral valve leaflets. *AoV* aortic valve, *Ah* annular height, *DAP* anterior-posterior diameter, *Syst* systole, *Diast* diastole

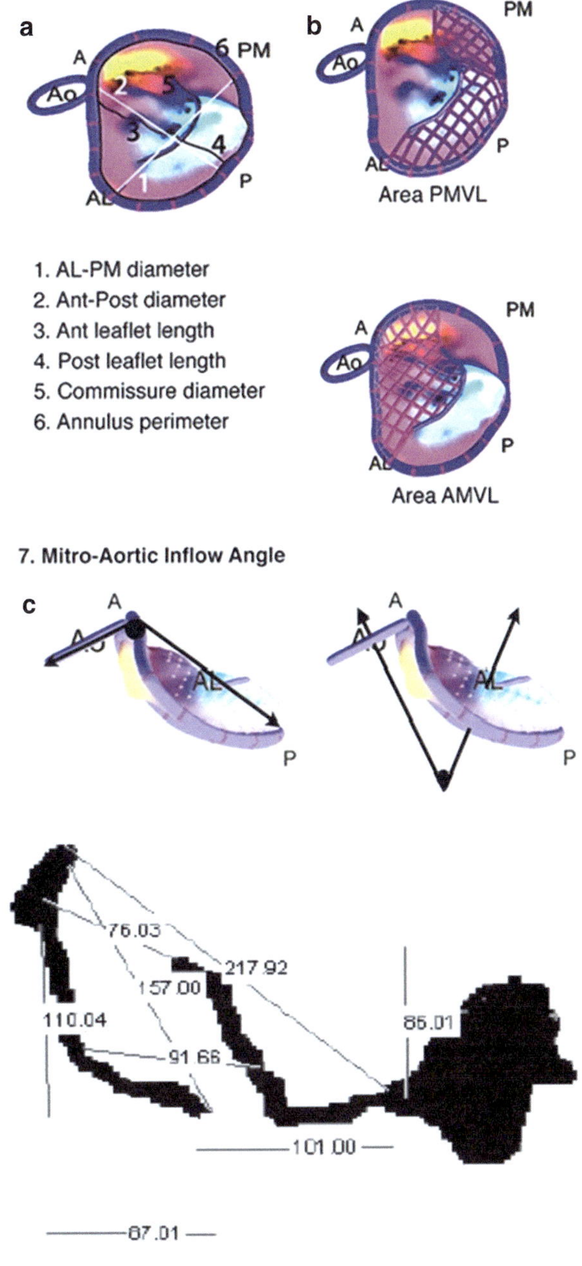

Fig. 10.4 The images show the quantitative evaluation of the mitral valve, which is shown with TomTec software. The images show the marking of the leaflets, which is done in several points of the cardiac cycle. The images in (**a**–**c**) are generated from the data set. From these, measures of annular dimensions, leaflet areas/lengths, and relations are obtained. The latter (**c**) include measure of the mitro-aortic angle (left: inflow; right: outflow) to help predict risk of systolic anterior motion of the mitral leaflets. *AL* anterolateral, *PM* posteromedial, *LA* left atrium, *LV* left ventricle, *Ao* aortic valve, *A* anterior, *P* posterior, *PMVL* posterior mitral valve, *AMVL* anterior mitral valve leaflet

Fig. 10.5 Gray lines provide vectors that attach bases, middles and tips of AML and PML together with MATLAB software

The resultant data potentially allow the implementation of an image-based approach for the patient-specific modeling of mitral valve leaflets. This approach could constitute the basis for the accurate evaluation of mitral valve pathologic conditions and for the planning of surgical approaches.

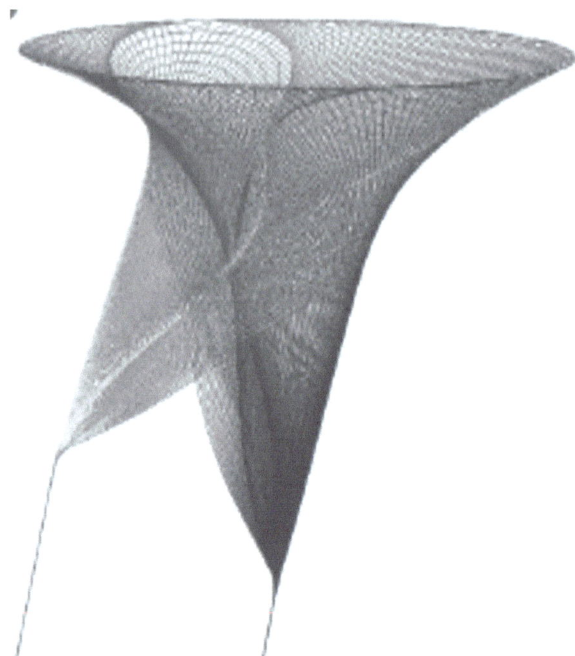

Fig. 10.6 3D AML and PML positions

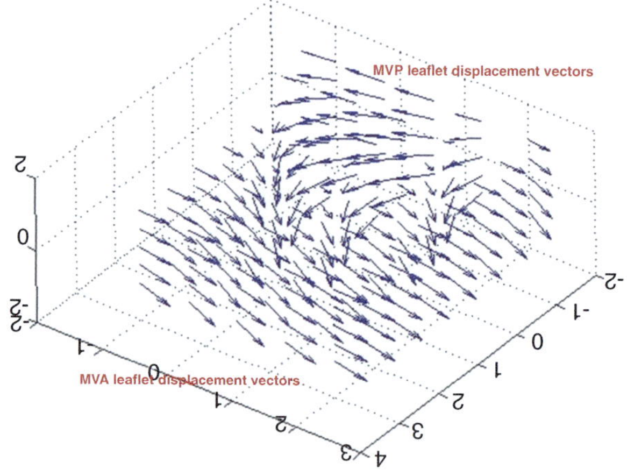

Fig. 10.7 Displacement vector assignments to mitral valve leaflets

Blood Forces Which Are Induced by Left Ventricular Muscular Forces

The local fluid velocity at a point of the elastic boundary is evaluated by interpolation from the Cartesian grid. The same approximate delta function that was used to spread force can also be used to get an interpolation operator that is the adjoint (or transpose) of the force spreading operators. In summary, the boundary method avoids many of the difficulties and pitfalls of the standard approach to fluid structure interaction. By representing an elastic boundary in terms of the forces applied by the immersed elastic boundary to the fluid, the boundary method avoids any assumption of boundary geometry in the fluid computation; makes it unnecessary to evaluate the fluid stress tensor at the elastic boundary; and makes it possible to simulate elastic boundaries, like the chambers of the human heart.

We assume that the fluid has uniform density, ρ, and uniform dynamic viscosity, μ. The structure is taken to be incompressible and neutrally buoyant, and the viscous properties of the structure are assumed to be those of the fluid in which it is immersed. Consequently, the momentum, velocity, and incompressibility of the coupled system can be described by the incompressible Navier-Stokes equations, augmented by an appropriately defined body force. (Even in the more complicated case in which the mass density of the structure differs from that of the fluid, the momentum, velocity, and incompressibility of the coupled system can still be described by the incompressible Navier-Stokes equations. The case in which the viscosity of the structure differs from that of the fluid can also presumably be done by a generalization of the methods proposed here, but this has not yet been attempted.)

The immersed boundary formulation of this problem employs an Eulerian description of the velocity and incompressibility of the fluid-structure system and a Lagrangian description of the configuration of the immersed elastic structure. In particular, the velocity of the entire coupled system is described in terms of an Eulerian velocity field, $u\ (x;\ t)$, where $x = (x;\ y;\ z)$ are fixed physical (Cartesian) coordinates 1, whereas the configuration of the immersed elastic structure is described in terms of a curvilinear coordinate system. Let $(q;\ r;\ s)$ be material curvilinear coordinates attached to the elastic structure so that fixed values of $(q;\ r;\ s)$ label a material point for all time t, with $X\ (q;\ r;\ s;\ t)$ referring to the Cartesian position of such a material point at time t. The physical domain consists of a region $U \subset R^3$. For simplicity, we presently take U to be the unit cube and impose periodic boundary conditions. The curvilinear coordinates are restricted to some region of $(q;\ r;\ s)$-space, here denoted $\Omega \subset R^3$. The configuration of the elastic structure at time t is denoted by $X\ (q, r, s, t)$, and the curvilinear force density (i.e., the density with respect to (q, r, s)) generated by the elasticity of the structure is determined by a time-dependent mapping from $X\ (q, r, s, t)$, the structure configuration at time t, to the elastic force density at time t, denoted $F\ (q, r, s, t)$.

$$f(x, t) = \int_\Omega F(q, r, s, t) \delta(x - X(q, r, s, t)) \, dq \, ds \, dr \qquad (10.1)$$

$f(x, t)$ is the (Cartesian) blood force density. The three-dimensional Dirac delta function, $\delta\ (x) = \delta\ (x)\ \delta\ (y)\ \delta\ (z)$, appears as the kernel of an integral transform that facilitates conversions between Eulerian and Lagrangian quantities. Equation (10.1) converts the curvilinear force density into the Cartesian force density. Note that their numerical values are generally not equal at corresponding points. Nevertheless, the Cartesian and curvilinear elastic force densities are equivalent as densities. Recalling the defining property of the Dirac delta function,

$$\int_V \delta\left(x-X\right) = \begin{cases} 1, & \text{if } X \epsilon V \\ 0, & \text{othewise,} \end{cases} \tag{10.2}$$

where $V \subset U$ is an arbitrary region of physical space.

Eulerian and Lagrangian Spatial Discretizations

The physical domain is discretized using a locally-refined Cartesian grid, but, for simplicity's sake, we describe only the uniform-grid version of this method. Let (i, j, k) index the Cartesian grid cells, and let $x_{i,j,k} = \left(i+\dfrac{1}{2}\right)h, \left(j+\dfrac{1}{2}\right)h, \left(k+\dfrac{1}{2}\right)h$ (Fig. 10.8) indicate the position of the center of grid cell (i, j, k), in which h is the Cartesian grid mesh width and $\Delta x = h^3\, x$ is the Cartesian grid cell volume. The Eulerian force densities $\mathbf{f}$ is approximated in the same staggered-grid fashion. The Eulerian pressure $\mathbf{p}$ is approximated at the centers of the Cartesian grid cells. The deformations and forces associated with the left ventricular volume sample are approximated on a fiber-aligned curvilinear mesh. Let l index the nodes of this mesh, let ϕ_l and F_l indicate the current position and the Lagrangian force density of node, respectively, and let Δq_l indicate the area fraction (quadrature weight) associated with node l. F_l is computed using a finite difference approximation to the fiber force density; see Chaps. 5 and 6 for details, including relevant finite difference formulae.

To couple the Lagrangian and Eulerian discretization, we employ approximation to integral transforms of continuum equations that replace the singular Dirac delta

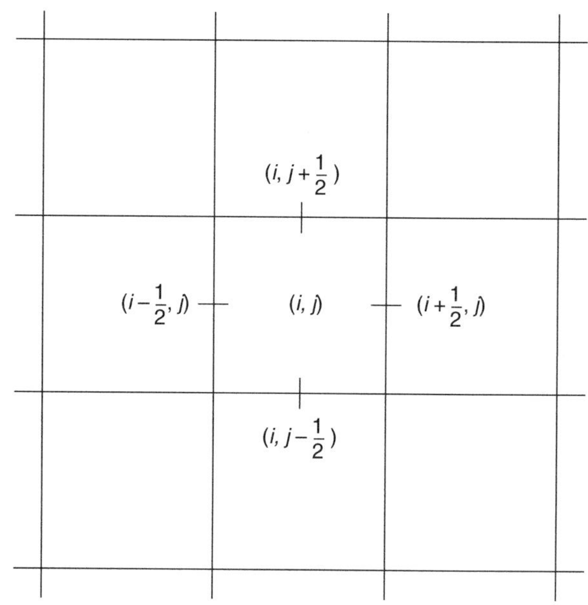

Fig. 10.8 Locations of cell-centered and face-centered quantities about Cartesian grid cell (i; j) for a two-dimensional grid. Placement on a three-dimensional grid is analogous

function $\delta(x) = \delta(x)\,\delta(y)\,\delta(z)$ by a regularized delta function $\delta_h(x) = \delta_h(x)\,\delta_h(y)\,\delta_h(z)$. In our computation, we construct the three-dimensional regularized delta either by using the four-point one dimensional regularized delta function, or by using a broadened version of this function that has a spatial extent of 8 mesh widths. The same regularized delta function is used for LV segments. The Lagrangian forces $F = (F^x, F^y, F^z)$ associated with the LV volume samples are converted into equivalent Eulerian forces (blood forces) $f = (f^x, f^y, f^z)$ via

$$f^x_{i-\frac{1}{2},j,k} = \sum_l F^x_l \delta_h\left(x_{i-\frac{1}{2},j,k} - \phi_l\right)\Delta q_l$$

$$f^y_{i,j-\frac{1}{2},k} = \sum_l F^y_l \delta_h\left(x_{i,j-\frac{1}{2},k} - \phi_l\right)\Delta q_l$$

$$f^z_{i,j,k-\frac{1}{2}} \sum_l F^z_l \delta_h\left(x_{i,j,k-\frac{1}{2}} - \phi_l\right)\Delta q_l$$

By the above formula and the elastic body forces of the LV muscles referred to in Chap. 6 and combining them with ST software to track the current position ϕ_l of node, we can calculate the blood forces which are induced by the left ventricular myocardial segments [19, 20].

References

1. Mittal R, Iaccarino G. Immersed boundary methods. Annu Rev Fluid Mech. 2005;37:239–61.
2. Kim WY, Walker PG, Pedersen EM, Poulsen JK, Oyre S, Houlind K. Left ventricular blood flow patterns in normal subjects: a quantitative analysis by three-dimensional magnetic resonance velocity mapping. J Am Soc Echocardiogr. 1995;26:24–38.
3. Kilner PJ, Yang G-Z, Wilkes AJ, Mohiaddin RH, Firmin DN, Yacoub MH. Asymmetric redirection of flow through the heart. Nature. 2000;404:59–61.
4. Bellhouse BJ. Fluid mechanics of a model mitral valve and left ventricle. Cardiovasc Res. 1972;6:199–210.
5. Domenichini F, Pedrizzetti G. Baccani, B.: three-dimensional filling flow into a model left ventricle. J Fluid Mech. 2005;539:79–98.
6. Kim HB, Hertzberg JR, Shandas R. Development and validation of echo PIV. Exp Fluids. 2004;36:55–62.
7. Zheng H, Liu L, Williams L, Hertzberg JR, Lanning C, Shandas R. Real time multicomponent echo particle image velocimetry technique for opaque flow imaging. Appl Phys Lett. 2006;88:1–3.
8. Sengupta PP, Khandheria BK, Korinek J, Jahangir A, Yoshifuku S, Milosevic I. Left ventricular isovolumic flow sequence during sinus and paced rhythms: new insights from use of high-resolution Doppler and ultrasonic digital particle imaging velocimetry. J Am Coll Cardiol. 2007;49:899–908.
9. Hong G-R, Pedrizzetti G, Tonti G, Li P, Wei Z, Kim JK, et al. Characterization and quantification of vortex flow in the human left ventricle by contrast echocardiography using vector particle image velocimetry. J Am Coll Cardiol Img. 2008;1:5–17.

10. Adrian RJ. Particle-imaging techniques for experimental fluid mechanics. Ann Rev Fluid Mech. 1991;23:261–304.
11. Kheradvar A, Kasalko J, Johnson D, Gharib M. An in vitro study of changing profile heights in mitral bioprostheses and their influence on flow. ASAIO J. 2006;52:34–8.
12. Batchelor GK. An introduction to fluid dynamics. Cambridge: Cambridge University Press; 1970.
13. Sengupta PP, Korinek J, Belohlavek M, Narula J, Vannan MA, Jahangir A. Left ventricular structure and function: basic science for cardiac imaging. J Am Coll Cardiol. 2006;48:1988–2001.
14. Karvandi M, Ranjbar S, Hassantash SA. Achievability of the assessment of the mitral valve leaflets by mathematical equations of Inelasticity based on echocardiography poster presentation at Americanecho. Abstract: P2–92; 2012.
15. Geu-Ru Hong GR, Pedrizzetti G, Giovanni Tonti G, Li P, Wei Z, Kim JK, Baweja A, Liu S, Namsik Chung N, Houle H, Narul J, Vannan MA. Characterization and quantification of vortex flow in the human left ventricle by contrast echocardiography using vector particle image velocimetry. JACC Cardiovasc Imaging. 2008;1:705–17.
16. Ranjbar S, Karvandi M, Hassantash SA. A novel mathematical based software for modeling the left ventricular myocardium. Poster presentation at Euroecho. Abstract; P448. 2012.
17. Ranjbar S, Karvandi M, Hassantash SA. A novel mathematical based software to demonstrate flow direction and curves inside the left ventricular myocardium. Poster presentation at Euroecho. Abstract P142; 2012.
18. Ranjbar S, Karvandi M, Hassantash SA. A novel mathematical fluid dynamic of left ventricular dilated cardiomyopathy. poster presentation at Americanecho. Abstract: P2–89; 2012.
19. Ranjbar S, Karvandi M, Hassantash SA. Impact of prosthetic mitral valve direction on blood flow path inside the left ventricle. Poster presentation at Euroecho. Abstract: P516; 2012.
20. Ranjbar S, Karvandi M, Ajzachi M. System and method modeling left ventricle of heart. US Patent. Patent number 8,414,490. 2013.

A Novel Quantitative Indicator of the Left Ventricular Contraction Based on Volume Changes of the Left Ventricular Myocardial Segments

11

Introduction

The systolic phase of the left ventricle is a complex process that comprises a coordinate contraction of subendocardial, midwall, and subepicardial muscle fibers. These fibers are arranged in a complex, helical manner. Midwall fibers are oriented circumferentially; the contraction of these fibers mostly contributes to a decrease in the minor axis of the left ventricle and is responsible for the generation of much of the ejection fraction (EF). Longitudinally oriented fibers in the subendocardium and the subepicardium contribute to a shortening of the long axis of the ventricle, which also contributesto EF. Resulting in the circumferential and longitudinal strains in normal cases, myocardial fiber movements begin from the posterior-basal region of the heart, continues through the LV free wall, touches the septum, rings around the apex, rises, and ends at the superior-anterior edge of LV [1, 2].

Contractile function is a general term to describe changes in performance or function that could be affected by changes in inotropic state, fiber length, or load. Thus, a change in the EF might best be classified as a change in contractile function; this term is often used interchangeably with the term systolic function. To fully describe the left ventricle it is necessary to define several additional terms that refer to its mechanical properties.

In a scenario involving this shortening and twisting deformation, wall thickening contributes to the volume displacement and creation of the EF. In the case of a LV myocardial segment (Fig. 11.1), the volume changes of it within a cardiac cycle, might be a quantitative marker of the regional function of the LV myocardium. Utilizing physical laws and mathematical equations, specific echocardiographic data can make available more detailed, valuable and practical information for the volume change measurements.

This study serves to provide a real-time parameter of regional LV contractility even more than one for global function of the left ventricular mechanisms.

© Springer Nature Switzerland AG 2023

M. Karvandi, S. Ranjbar, *A Review on Recent Echocardiographic Software*,

https://doi.org/10.1007/978-3-031-29046-6_11

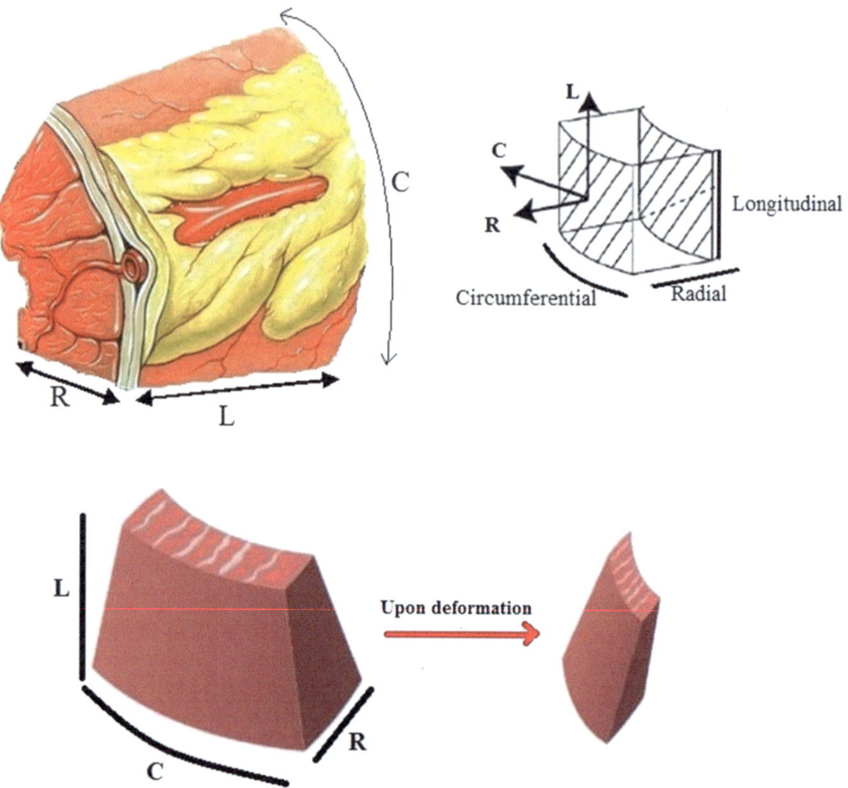

Fig. 11.1 An anatomical LV myocardial sample (from the top at the left side), A local coordinate attached to it (from the top at the right side), A LV myocardial segment upon deformation (the bottom). *L* longitudinal direction, *R* radial direction, *C* circumferential direction

Methods and Results

Patients who underwent clinically-directed standard transthoracic echocardiography using 2D conventional echocardiography machines calibrated to measure strain components (longitudinal, radial and circumferential strains) were asked to acquire datasets from apical 4-chamber (4-C), 2-chamber (2-C) and short axis views. An expert investigator was requested to first qualitatively estimate displacements and longitudinal, radial and circumferential strains for each LV echocardiographic segments per cardiac cycle (Fig. 11.2).

Volume fractional changes of the LV echocardiographic segments are expanded based on their strains indices over time. For a muscle myocardial sample of the left ventricle, we named its volume by $V(t)$ at time t.

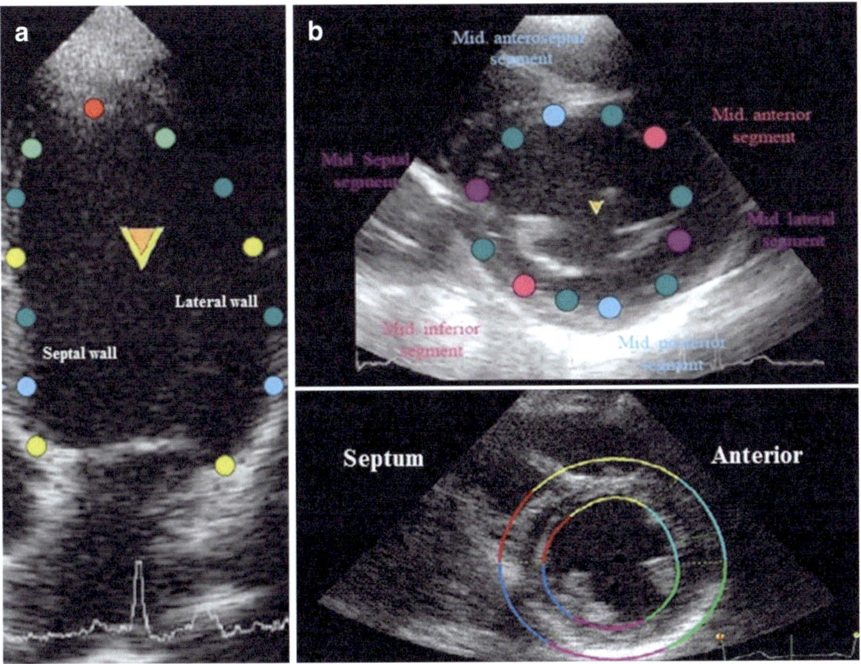

Fig. 11.2 Illustrations showing a myocardial segment model of the left ventricular wall in different views (**a, b**)

$$\frac{\nabla V(t)}{V(t_\mathrm{d})} = \frac{\left[V(t) - V(t_\mathrm{d}) \right]}{V(t_\mathrm{d})}$$ which t_d is the initial time at the end diastole.

$V(t) = L(t)R(t)C(t)$ and $V(t_\mathrm{d}) = L(t_\mathrm{d})R(t_\mathrm{d})C(t_\mathrm{d})$. Let

$$\varepsilon_L(t) = \frac{\left[L(t) - L(t_d) \right]}{L(t_d)}; \varepsilon_R(t) = \frac{\left[R(t) - R(t_d) \right]}{R(t_d)}; \varepsilon_C(t) = \frac{\left[C(t) - C(t_d) \right]}{C(t_d)}$$ are longitudinal strain, radial strain and circumferential strain, respectively, at time t. We can easily rewrite the $\dfrac{\nabla V(t)}{V(t_\mathrm{d})}$ in terms of strain components by the following way:

$$L(t) = L(t_\mathrm{d})\varepsilon_L(t) + L(t_\mathrm{d}); R(t) = R(t_\mathrm{d})\varepsilon_R(t) + R(t_\mathrm{d}); C(t) = C(t_\mathrm{d})\varepsilon_C(t) + C(t_\mathrm{d})$$

$$\frac{\nabla V(t)}{V(t_\mathrm{d})} = \frac{\left[V(t) - V(t_\mathrm{d}) \right]}{V(t_d)} = \frac{\left[L(t)R(t)C(t) - L(t_\mathrm{d})R(t_\mathrm{d})C(t_\mathrm{d}) \right]}{L(t_\mathrm{d})R(t_\mathrm{d})C(t_\mathrm{d})}$$

$$\frac{\nabla V(t)}{V(t_0)} = \left[\left(L(t_\mathrm{d})\varepsilon_L(t) + L(t_\mathrm{d}) \right) \times \left(R(t_\mathrm{d})\varepsilon_R(t) + R(t_\mathrm{d}) \right) \right.$$
$$\left. \times C(t_\mathrm{d})\varepsilon_C(t) + C(t_\mathrm{d}) \right) / L(t_\mathrm{d})R(t_\mathrm{d})C(t_\mathrm{d}) \right] / L(t_d)R(t_d)C(t_d)$$
$$= \varepsilon_L(t)\varepsilon_R(t)\varepsilon_C(t) + \varepsilon_L(t)\varepsilon_R(t) + \varepsilon_L(t)\varepsilon_C(t) + \varepsilon_R(t)\varepsilon_C(t)$$
$$+ \varepsilon_L(t) + \varepsilon_R(t) + \varepsilon_C(t)$$

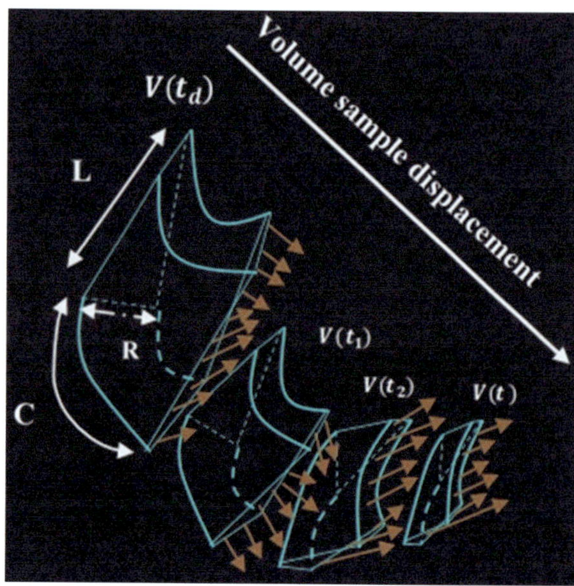

Fig. 11.3 Brown vectors represent the impact of the myocardial volume changes to shift the blood fluid in the LV cavity within a cardiac sample

$$\frac{\nabla V(t)}{V(t_0)} = \varepsilon_L(t)\varepsilon_R(t)\varepsilon_C(t) + \varepsilon_L(t)\varepsilon_R(t) + \varepsilon_L(t)\varepsilon_C(t) + \varepsilon_R(t)\varepsilon_C(t) + \varepsilon_L(t)$$
$$+ \varepsilon_R(t) + \varepsilon_C(t)$$

The above formula states the volume fractional changes as a function of longitudinal, radial and circumference strains. Let **DV**(*t*) be the displacement of the volume *V*(*t*) **at time** *t* using thermodynamics laws in physics (Fig. 11.3), the ejected blood volume fraction at time *t* induced by $\nabla V(t)$ is $\dfrac{\nabla V(t)}{V(t_d)} \times \mathbf{DV}(t)$ at the time *t*.

Finally, the total ejected blood volume fraction (TEBVF) by the myocardial segments in the left ventricular cavity is obtained by the integration from $\dfrac{\nabla V(t)}{V(t_d)} \times \mathbf{DV}(t)$ over the times and LV myocardial segments.

$$\mathbf{TEBVF}(t) = \int_{t_d}^{t}\int_{V\,\text{runs through the LV myocaldial samples}} \frac{\nabla V(t)}{V(t_d)} \times \mathbf{DV}(t)$$

$$= \int_{t_d}^{t}\int_{V\,\text{runs through the LV myocaldial samples}} \left[\varepsilon_L(t)\varepsilon_R(t)\varepsilon_C(t) + \varepsilon_L(t)\varepsilon_R(t) + \varepsilon_L(t)\varepsilon_C(t)\right.$$
$$\left. + \varepsilon_R(t)\varepsilon_C(t) + \varepsilon_L(t) + \varepsilon_R(t) + \varepsilon_C(t)\right] \times \mathbf{DV}(t)$$

Such that $t \in [t_d, t_s]$ and t_s is the time at the end systolic phase.

Discussion

1. The new introduced volume strain $VS(t) = \dfrac{\nabla V(t)}{V(t_d)}$ of a myocardial segment of the left ventricle over time is a regional computable indicator based on strain components of the LV mechanism. And the global volume strain $\mathbf{GVS}(t) = \sum_V \dfrac{\nabla V(t)}{V(t_d)}$ (where V runs through all LV's volume samples) is a parameter to fully describe the global LV mechanic at time t. $\mathbf{TVS}(t) = \displaystyle\int_{t_d}^{t} \sum_V \dfrac{\nabla \mathbf{V}(t)}{\mathbf{V}(t_d)}$ is the total volume strain of the left ventricular myocardium over the time.

2. We obtained several left ventricular contractility variables

$$\mathbf{LVC}(t) = \frac{\nabla \mathbf{V}(t)}{\mathbf{V}(t_d)} \times \mathbf{DV}(t), \qquad \mathbf{GLVC}(t) = \sum_V \frac{\nabla V(t)}{V(t_d)} \times \mathbf{DV}(t) \qquad \textbf{and}$$

$$\mathit{TE}\mathbf{BVF}(t) = \int_{t_d}^{t} \sum_V \frac{\nabla V(t)}{V(t_d)} \times \mathbf{DV}(t) = \int_{t_d}^{t} \sum_V \varepsilon_L(t)\varepsilon_R(t)\varepsilon_C(t) + \varepsilon_L(t)\varepsilon_R(t)$$
$$+ \varepsilon_L(t)\varepsilon_C(t) + \varepsilon_R(t)\varepsilon_C(t) + \varepsilon_L(t) + \varepsilon_R(t) + \varepsilon_C(t) \times DV(t)$$

in the LV cavity based on the volume strains, the global volume strains and the total volume strains of the left ventricular myocardial samples.

3. TEBVF is realized as a function such that the EF has an especial value $(\mathrm{EF} = \mathrm{TEBVF}(t_s))$.

4. TEBVF is an independent parameter of the heart load and heart tethering and it can be estimated automatically by the formula.

5. It may introduce a non-invasive method to test the viability in patients with ischemic heart diseases.

6. As an application, TEBVF (t_s) gives the real value of EF and this value has a main role to make an exact decision before surgical tasks. For a patient with sever mitral valve regurgitation and EF 40%, a MVR or a mitral valve repair is reported for a cardiac surgeon. Since a fraction of the blood volume was regurgitated to the left atrium, TEBVF (t_s) with the value 10% makes the patient a candidate for a heart transplantation. For another example, in the case of a patient with the aortic valve stenosis and high blood pressure toward the left ventricular hypertrophy it might be possible to use $\mathrm{TEBVF}(t_s)$ for the real value of EF.

7. According to the continuous geometry [3], a 3D reconstructive image of the LV from conventional 2D echocardiographic images makes a border tracking of the LV. First, all 2D echocardiographic images in all different views are acquired from a 2D echocardiography machine; then region of interests as the considered segments would be determined by the speckle tracking software [4]. A range of the mesh screens can be achieved alongside these images by using the 4D LV-analysis function software (TomTec Imaging Systems GmbH, Munich, Germany)

in which each network includes a LV segment. These networks (meshes) are connected together and a 3D mathematical model of the left ventricle would be created. Since this 3D LV reconstruction is based on the manipulation of 2D images, our formulas can be easily applied to this 3D LV modeling to gain as one of the exact approaches for estimations of introduced variables.

8. Christian Knackstedt et al. [5] introduced a fully automated versus a standard tracking of left ventricular (LV) ejection fraction (EF) and global longitudinal strain (GLS), using the machine learning-enabled image analysis (AutoLV, TomTec-Arena 1.2, TomTec Imaging Systems, Unterschleissheim, Germany). Voigt et al. [4] and Farsalinos et al. [6] and Thomas H. Marwick et al. [7] have recently also studied the newer algorithms that rely primarily on speckle tracking and artificial intelligence in tracking techniques to detect and contour the endocardium and cardiac cycle and they run the LS, EF and analogous measurements. One questionable concern that arises as a result of these studies is whether or not this progress in terms of the measurement of strain will allow GLS to displace EF. Although GLS is the strongest of the deformation markers, reproducing the promising effect of averaging when individual measurements are subject to noise, we need a computable indicator of regional function even more than one for global function. The new provided quantitative marker (TEBVF(t)) in this chapter suggests the same needed calculable indicator be used in the analysis the left ventricular function.

Conclusion

The TEBVF parameter has important implications to give the real value of EF in the sever Mitral valve regurgitations and makes the lack of advance and expensive echocardiographic software and will allow TEBVF to replace GLS and EF.

References

1. Kocica MJ, Corno AF, Carreras-Costa F, Ballester-Rodes M, Moghbel MC, Cueva CNC, Lackovic V, Kanjuh VI, Torrent-Guasp F. The helical ventricular myocardial band: global, three-dimensional, functional architecture of the ventricular myocardium. Eur J Cardiothorac Surg. 2006;29:S21–40.
2. 4D LV-Analysis-TomTec–Imaging Systems. http://www.tomtec.de/end_users/4d_echo/4d_lv_analysisc.html. Accessed Mar 2015.
3. Neumann JV. Continuous geometry, vol. 22. Princeton: Princeton University Press; 1960. p. 92–100.
4. Voigt JU, Lysyansky P, Marwick TH, Houle H, Baumann R, et al. Definitions for a common standard for 2D speckle tracking echocardiography: consensus document of the EACVI/ASE/industry task force. J Am Soc Echocardiogr. 2015;28:183–93.
5. Knackstedt C, Bekkers SCAM, Schummers G, et al. Fully automated versus standard tracking of left ventricular ejection fraction and longitudinal strain. J Am Coll Cardiol. 2015;66:1456–66.
6. Farsalinos KE, Daraban AM, Ulnu S, Thomas JD, Badano LP, Voigt JU. Head-to-head comparison of global longitudinal strain measurements among nine different vendors the EACVI/ASE inter-vendor comparison study. J Am Soc Echocardiogr. 2015;28:1171–81.
7. Marwick TH. The clinical application of strain: raising the standard. J Am Soc Echocardiogr. 2015;28:1182–3.

Different Mathematical Techniques to Measure Left Ventricular 2D Deformations: Strain Imaging

12

Different Mathematical Algorithms to Quantify LV 2D Strains

1. Tissue Doppler imaging algorithm based on ultrasound method combined with velocity gradient (TDI-VG) [1]:
   (a) Echocardiographic images are acquired from LV short axis and apical views.
   (b) The region of interest L_1 in the left ventricular myocardium
   (c) is manually tracked with velocity V_1
   (d) After passing time t the search region L_2 is detected by Speckle tracking software with velocity V_2 (Fig. 12.1).
   (e) Let L be the distance between L_1 and L_2 then strain rate (ε') of L_1 at time t is formulated using TDI-VG method by the following way:

$$\varepsilon'\left(L_1,t\right) = \frac{V_2\left(t\right)-V_1}{L\left(t\right)}$$

   (f) The strain magnitude of the region of interest L_1 at time t is:

$$\varepsilon\left(L_1,t\right) = \int_0^t \varepsilon'\left(L_1,u\right)\mathrm{d}u$$

2. Speckle tracking combined with end-7 points algorithm (ST-end 7 points) [2–6]:
   (a) Echocardiographic images are acquired from LV short axis and apical views.
   (b) Seven known constant points are automatically traced in the left ventricular myocardium by speckle tracking software (Fig. 12.2).
   (c) j_n; Search regions of the ith point at time t_n in the n-th frame (Fig. 12.3).
   (d) L_{n,t_n}'s ; Distances between ith point (first frame) and j_n corresponding to time t_n.
   (e) Strain value of the ith point at time t_n :

$$\varepsilon\left(i,t_n\right) = \sum_{k=1}^{n-1}\left(L_{k+1,t_{k+1}} - L_{k,t_k}\right)/L_{k,\,t_k}$$

© Springer Nature Switzerland AG 2023

M. Karvandi, S. Ranjbar, *A Review on Recent Echocardiographic Software*,

https://doi.org/10.1007/978-3-031-29046-6_12

Fig. 12.1 The TDI-VG based method to obtain strain by time integrating the strain rate curve. V_1, velocity at the region of interest 1; V_2, velocity at the search region 2; L, distance between 1 and 2 regions

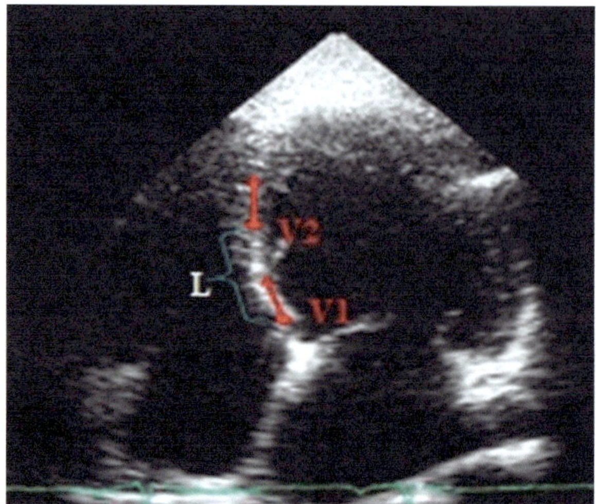

Fig. 12.2 The ST-end 7 points-based method to obtain strain by displacement tracking of 7 points (1–7) per cardiac cycle

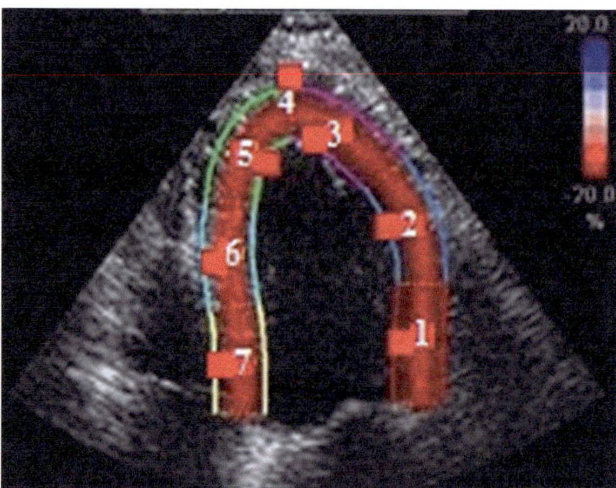

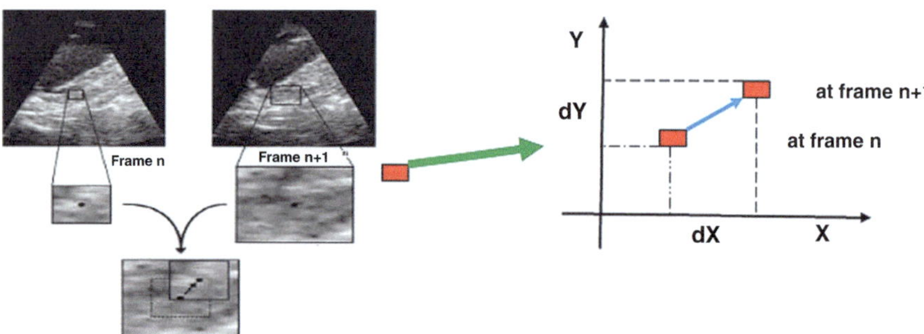

Fig. 12.3 The only Speckle tracking method to obtain strain by searching the region of interest frame by frame through a single cardiac cycle. dX speckle displacement from frame n to frame $n + 1$ along the X-axis, dY speckle displacement from frame n to frame $n + 1$ along the Y-axis

3. Speckle tracking combined with TDI algorithm [7, 8].
   (a) Echocardiographic images are acquired from LV short axis and apical views.
   (b) Let A is an arbitrary region of interest which is traced manually.
   (c) B to N, search points that are automatically tracked by the Speckle tracking software.
   (d) L_n, the change of length from point A to N in the n-th frame.
   (e) $T_i's$, time corresponding to the search points.
   (f) $L_{AN,n}$, the specified area corresponding to N at TDI velocity gradient diagram in the n-th frame.
   (g) Strain value of the point A at time T_i is measurable by the following formula: (Fig. 12.4)

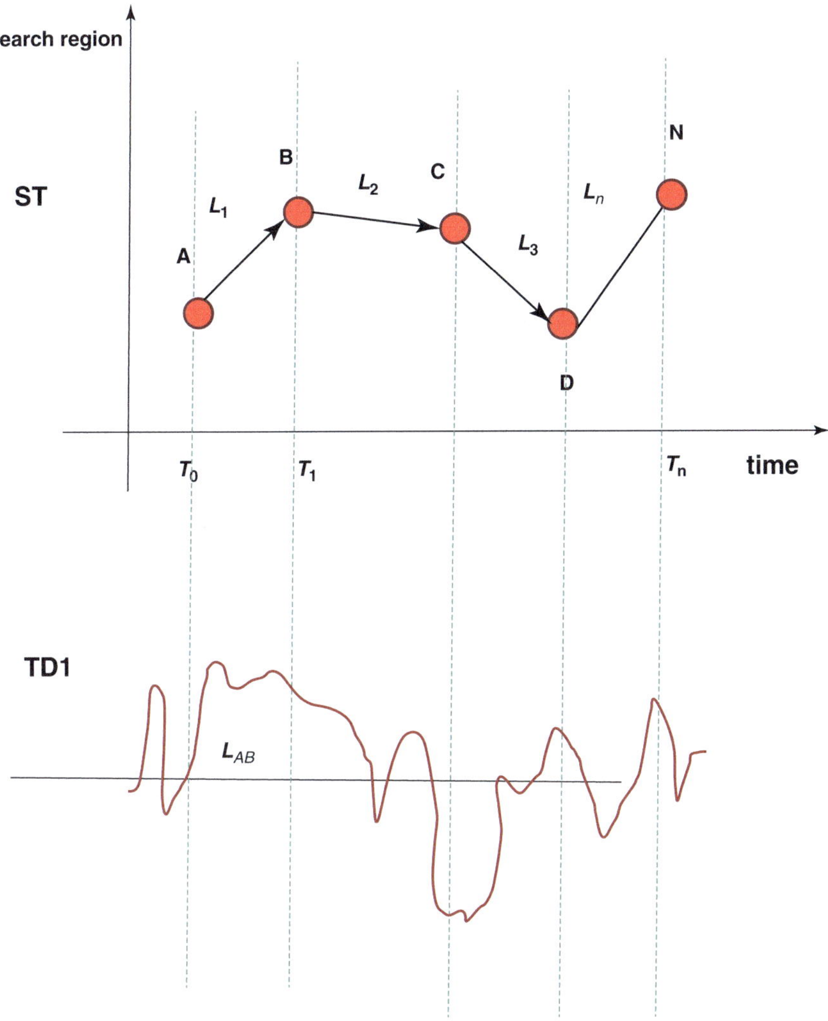

Fig. 12.4 The ST-TDI-based method to obtain strain by combining Speckle displacements to TDI displacements. *A* region of interest, *B* to *N* search points, $T_i's$ time corresponding to each search points, L_1 to L_n the changes of lengths from point *A* to *N*, L_{AB} the specified area corresponding to L_1

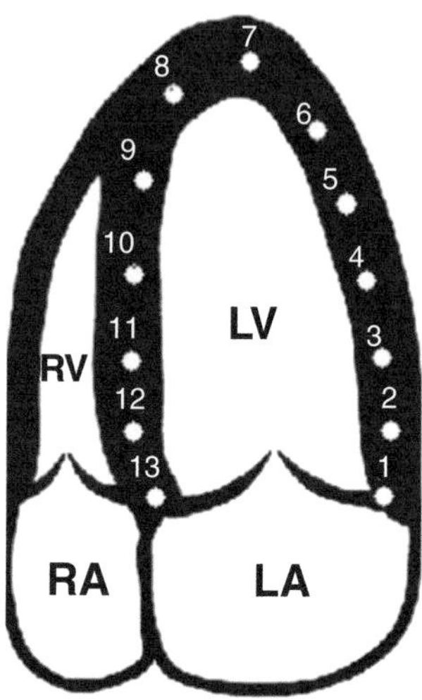

Fig. 12.5 LV myocardial region has been divided to 13 points at a 4C view

$$\varepsilon\left(A,T_i\right)=\sum_{n=1}^{i}\left(L_{AN,n}\left(T_i\right)-L_{A(N-1),n-1}\left(T_{i-1}\right)\right)/L_{A(N-1),n-1}\left(T_{i-1}\right)$$

4. Velocity vector imaging algorithm (VVI) [9]:
   (a) Echocardiographic images are acquired from LV 4 apical chamber (4C) views.
   (b) $A_{1,k}$'s; Regions of interests of the left ventricular myocardium that are manually divided to 13 points at 4C (first frame); $k = 1,2,3,..,13$ (Fig. 12.5).
   (c) Considering perpendicular lines crossing each divided points ($k = 1,2,3,...,13$) (Fig. 12.6).
   (d) $A_{2,k}$; New points are automatically identified from the intersection between perpendicular lines and the second frame ($k = 1,2,3...,13$).
   (e) $A_{i,k}$; New regions by the same procedures at last steps for the ith frame from 1-th frame to ith frame).
   (f) T_i; Time corresponding to ith frame.
   (g) $l_{i+1,T_i,k}$; Distances between $A_{i+1,k}$ and $A_{i,k}$ corresponding to T_i for each $k = 1,2,3,...,13$.
   (h) $\varepsilon\left(A_{1,k},T_{i+1}\right)=\sum_{j=1}^{j=i+1} l_{j+1,T_{j+1},k}-l_{j,T_j,k}/l_{j,T_j,k}$; Strain value of points $A_{1,k}$ for each k at time T_{i+1}.

5. Speckle tracking combined with LV as a polygon shape (a 4D = 3D + 1D strain which its 3D is due to left ventricle as the polygon modeling and 1D is over the time) algorithm [10].
   (a) Echocardiographic images are acquired from LV short axis and apical views.
   (b) The regions of left ventricular myocardium are divided into 7 points and 4 points at 4C and SX views, respectively.

Fig. 12.6 Perpendicular lines crossing each 13 points

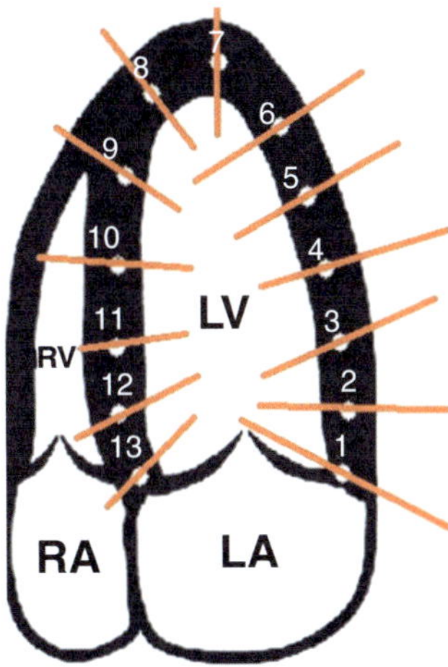

Fig. 12.7 Red points are corresponding geometrical points to material points of the LV myocardium. Material points are usually selected as contact points in the myocardium

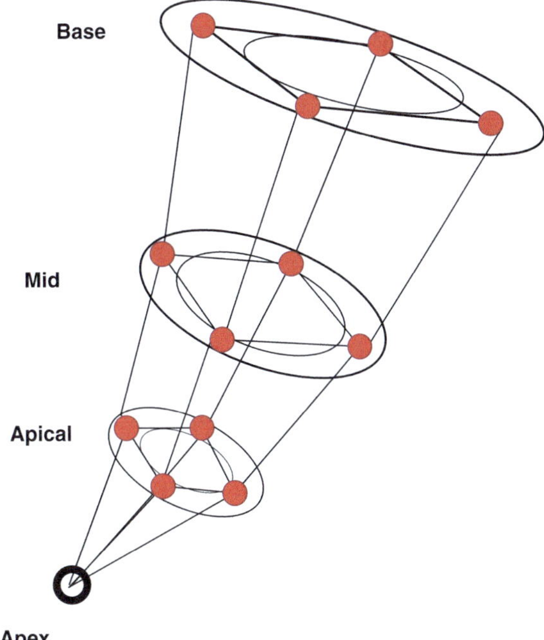

(c) Myocardial point connections together can be realized by a polygon (Fig. 12.7).

(d) Speckle tracking method applied to reconstructed LV as a polygon (Fig. 12.8).

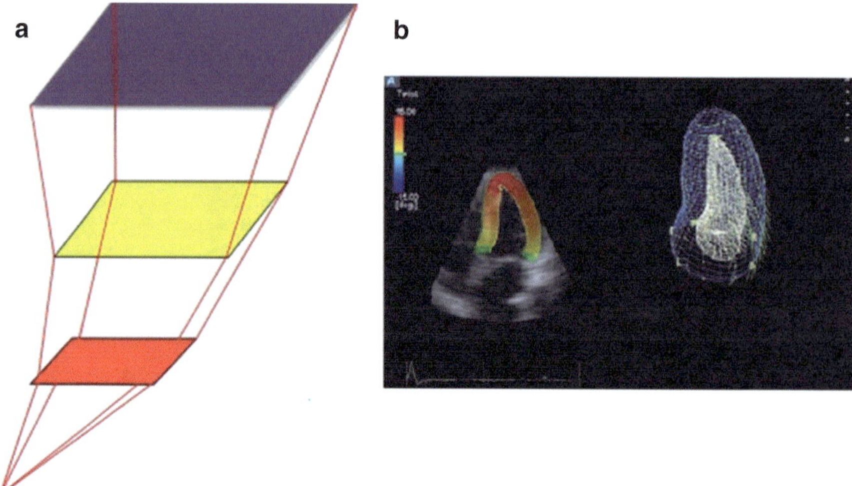

Fig. 12.8 (**a**) The left ventricle as a polygon shape modeling. (**b**) Speckle tracking method applied to LV polygon shape. *LV* left ventricle

Different Ultrasound Machines Applied to Computations of Left Ventricular Global Strains

Different echocardiography systems can measure LV strain using above mathematical algorithms (1–5) [11–13].

E1 = Vivid 7, GE Healthcare system (TDI-VG).
E2 = Vivid E9, GE Healthcare system (ST-end 7 points).
E3 = iE33, Philips Medical system (ST-TDI).
E4 = Siemens Medical system (ST-VVI).
E5 = Artida or Aplio, Toshiba Medical Systems (ST-LV as a polygon shape).

The Global Strain Values

We have reviewed explicitly the left ventricular 2D strain for each different type of strain methods from five echocardiography machines. This manuscript introduces a mention value for strain, which is worthwhile in future basic and clinical research by the five specified software packages (Table 12.1).

Table 12.1 Left ventricular global strain components at five different modern echocardiography systems

	Method	GLLS	GLRS	GLCS
E1	TDI-VG	−18.33 ± 5.66%	36.6 ± 10.2%	−21.5 ± 5.7%
P-value		0.006	0.006	0.006
E2	ST- end 7 points	−21.3 ± 2.1%	54.6 ± 12.6%	−22.8 ± 2.9%
P-value		<0.0001	<0.0001	<0.0001
E3	ST-TDI	−18.9 ± 2.5%	36.3 ± 8.2%	−22.2 ± 3.2%
P-value		<0.0001	<0.0001	<0.0001
E4	ST-VVI	−20.66 ± 4.0%	52.1 ± 9.6%	−31.11 ± 5.8%
P-value		0.02	0.02	0.02
E5	ST-LV as a polygon shape	−19.9 ± 2.4%	51.4 ± 8.0%	−30.5 ± 3.8%
P-value		<0.0001	<0.0001	<0.0001

GLLS global longitudinal strain, *GLRS* global radial strain, *GLCS* global circumferential strain

References

1. Heimdal A, Stoylen A, Torp H, Skjaerpe T. Real-time strain rate imaging of the left ventricle by ultrasound. J Am Soc Echocardiogr. 1998;11:3–9.
2. Amundsen BH, Crosby J, Steen PA, Torp H, Slørdahl SA, Støylen A. Regional myocardial long-axis strain and strain rate measured by different tissue Doppler and speckle tracking echocardiography methods: a comparison with tagged magnetic resonance imaging. Eur J Echocardiogr. 2009;10:229–37.
3. Amundsen BH, Helle-Valle T, Edvardsen T, Torp H, Crosby J, Lyseggen E, et al. Noninvasive myocardial strain measurement by speckle tracking echocardiography: validation against sonomicrometry and tagged magnetic resonance imaging. J Am Coll Cardiol. 2006;47:89–93.
4. Langeland S, Dohooge J, Wouters PF, Leather HA, Claus P, Bijnens B, et al. Experimental validation of a new ultrasound method for the simultaneous assessment of radial and longitudinal myocardial deformation independent of insonation angle. Circulation. 2005;112:57–62.
5. Rappaport D, Adam D, Lysyansky P, Riesner S. Assessment of myocardial regional strain and strain rate by tissue tracking in B-mode echocardiograms. Ultrasound Med Biol. 2006;32:81–92.
6. Toyoda T, Baba H, Akasaka T, Akiyama M, Neishi Y, Tomita J, et al. Assessment of regional myocardial strain by a novel automated tracking system from digital image files. J Am Soc Echocardiogr. 2004;17:4–8.
7. Cho GY, Chan J, Leano R, Strudwick M, Marwick TH. Comparison of two dimensional speckle and tissue velocity based strain and validation with harmonic phase magnetic resonance imaging. Am J Cardiol. 2006;97:1–6.
8. Hanekom L, Cho GY, Leano R, Jeffriess L, Marwick TH. Comparison of two-dimensional speckle and tissue Doppler strain measurement during dobutamine stress echocardiography: an angiographic correlation. Eur Heart J. 2007;28:65–72.
9. Avi VM, Lang RM, Badano LP, Belohlavek M, Cardim NM, Derumeaux G, et al. Current and evolving echocardiographic techniques for the quantitative evaluation of cardiac mechanics: ASE/EAE consensus statement on methodology and indications endorsed by the Japanese Society of Echocardiography. Eur J Echocardiogr. 2011;12:167–205.

10. Orderud F, Kiss G, Langeland S, Remme E, Torp HG, Rabben SI. Combining edge detection with speckle-tracking for cardiac strain assessment in 3D echocardiography. In: Ultrasonics symposium. IUS 2008. Beijing: IEEE; 2008. p. 1959–62. https://doi.org/10.1109/ULTSYM.2008.0483.

11. Geyer H, Caracciolo G, Abe H, Wilansky S, Carerj S, Gentile F, et al. Assessment of myocardial mechanics using speckle tracking echocardiography: fundamentals and clinical applications. J Am Soc Echocardiogr. 2010;23:51–69.

12. Mondillo S, Galderisi M, Mele D, Cameli M, Lomoriello VS, Zaca V, et al. Speckle tracking echocardiography: a new technique for assessing myocardial function. J Ultrasound Med. 2011;30:71–83.

13. Voigt JU, Pedrizzetti G, Lysyansky P, Marwick TH, Houle H, Baumann R, et al. Definitions for a common standard for 2D speckle tracking echocardiography: consensus document of the EACVI/ASE/industry task force. J Am Soc Echocardiogr. 2015;28:183–93.

Mathematical Bernoulli's Equation in the Mitral Valve of Heart-Based Fractional Differentiations

13

Background

In the past decade, finite element modeling of complete heart valves has greatly aided evaluation of heart valve surgery, the design of bioprosthetic valve replacements, and the general understanding of healthy and abnormal cardiac function. Such a model must be based on an accurate description of the mechanical behavior of the valve material. It is essential to calculate velocity/displacement, geometrical indices and strain rate/strain at a component level that is to work at the cellular level. In this study we developed the first three-dimensional mitral valve modeling based on Bernoulli's equation and mathematical techniques to solve them in the characterization of mitral valve leaflets in continuum equations of inelasticity framework based on echocardiography.

We suggest a problem. A fluid reservoir of circular symmetry is to be designed. The minimum and maximum heights of the reservoir above the ground are h and H, respectively. The fluid exits through the mitral valve orifice at H. The time for the fluid to reach a particular height above the ground, say height z, is given by $t(z)$ where $t(H) = 0$. Determine the shape of mitral valve leaflets. The objective of this chapter is to study and give an algorithm for the mechanisms of the mitral valve of the heart grounded in mathematical techniques and Bernoulli's equation based on echocardiographic datasets.

Method and Results

The problem is to determine the shape $f(y)$ of a wire equation notch (here the notch is considered the orifice of the mitral valve and a wire equation is realized by the streamline flow behind the mitral valve and a modeling of mitral valve leaflets), an opening in mitral valve leaflets, in which the volume flow rate of fluid, Q, through the orifice is expressed as function of height h of the orifice [1]. We first establish the equation.

© Springer Nature Switzerland AG 2023

M. Karvandi, S. Ranjbar, *A Review on Recent Echocardiographic Software*,

https://doi.org/10.1007/978-3-031-29046-6_13

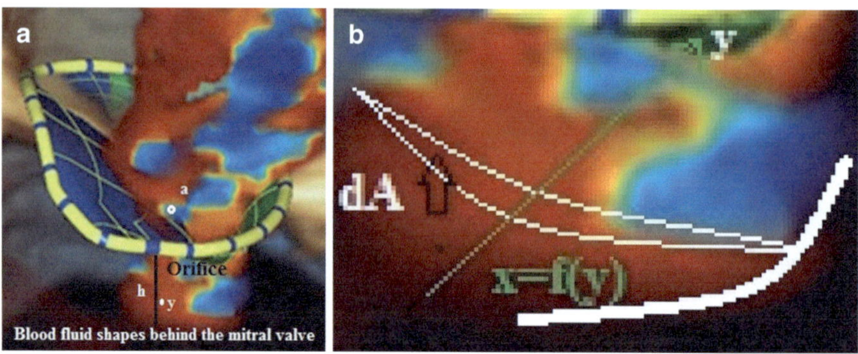

Fig. 13.1 Consider front (**a**) and side (**b**) views of the orifice of the mitral valve

$$Q(h) = c\int_0^h (h-y)^{\frac{1}{2}} f(y)\,\mathrm{d}y.$$

Assuming Bernoulli's equation can be applied between points "a" and "y" in Fig. 13.1 [2], we obtain

$$(*)\frac{P_a}{\rho} + gh + \frac{V_a^2}{2} = \frac{P_y}{\rho} + gy + \frac{V_y^2}{2}.$$

where P_a, P_y, V_a, V_y are the pressures and velocities at points "a" and "y"; g is the gravitational acceleration and ρ is the fluid density.

The pressures at "a" and "y" are both taken to be nearly atmospheric so that $P_a = P_y$ and the velocity at "a" is assumed to be negligible ($V_a = 0$) since the fluid behind the orifice is slow-moving. Thus, (*) becomes

$$gh = gy + \frac{V_y^2}{2}$$

So that

$$V_y = (2g)^{1/2}(h-y)^{1/2}$$

Gives the velocity of the fluid at distance "y" behind the orifice of the mitral valve in Fig. 13.1.

The elemental area (shaded region in Fig. 13.1b) is given by:

$$dA = 2f(y)\,\mathrm{d}y.$$

So, by definition, the elemental volume flow rate through dA is

$$(**)dQ = V_y\,dA = 2(2g)^{\frac{1}{2}}(h-y)^{\frac{1}{2}} f(y)\,\mathrm{d}y.$$

Denoting $2(2g)^{\frac{1}{2}}$ by c and integrating (**) from $y = 0$ to $y = h$ gives the total volume flow rate through the orifice of the mitral valve:

$$Q(h) = c\int_0^h (h-y)^{\frac{1}{2}} f(y)\,dy$$

We find $f(y)$ by finding $f(h)$ inversely by the following formula:

$$f(h) = 1/\Gamma\left(\frac{3}{2}\right) \times \frac{d^2}{dh^2} \times 1/\Gamma\left(\frac{1}{2}\right) \times \int_0^h (h-y)^{-\frac{1}{2}} g(y)\,dy$$

where $\dfrac{Q(h)}{c} = g(h)$ and Γ is the Gamma function. Since $\dfrac{Q(h)}{c}$ is known, $g(h)$ and $g(y)$ are known. For the mitral valve with numerical calculations of the leaflets in echocardiography machines, we can consider $g(h) = h^a$ approximately. For example for $a = 7/2$, the blood fluid is shaped like a parabola for the normal mitral valve of the heart. So by these shapes we can obtain a geometrical modeling which are coded and run in the MATLAB software (Figs. 13.2 and 13.3).

In fact, the above formula is a solution of the Lagrange–Euler equations applied around the mitral valve based on echocardiographic datasets:

$$\frac{\partial L\left(P_{i,}\,\dot{P}_i,\varepsilon_P,\dot{\varepsilon}_P,t\right)}{\partial P_i} - \frac{d}{dt}\left(\frac{\partial L\left(P_i,\dot{P}_i,\varepsilon_{P,i},\dot{\varepsilon}_{P,i,t}\right)}{\dot{P}_i}\right) = F\left(P_i,\dot{P}_i,\varepsilon_{P,i},\dot{\varepsilon}_{P,i},t\right)$$

$$\begin{array}{ccc} \varepsilon_{P,1} & \varepsilon_{P,2} & \varepsilon_{P,3} \\[4pt] \varepsilon_{P,4} & \varepsilon_{P,5} & \varepsilon_{P,6} \\[4pt] \varepsilon_{P,7} & \varepsilon_{P,8} & \varepsilon_{P,9} \end{array}$$

$P_1, P_2,\ldots,P_9$ are nine position variables toward nine strain components $\varepsilon_{P,1}$, $\varepsilon_{P,2}$, $\varepsilon_{P,3},\ldots$, $\varepsilon_{P,9}$ that are made by the Speckle tracking method on echocardiographic images. L is the Lagrangian, which is kinetic energy, and mines the potential energy of our system at the vicinity of the mitral valve.

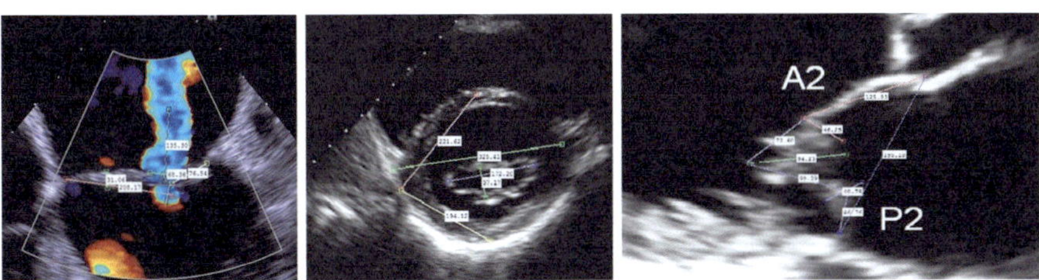

Fig. 13.2 Mitral valve measurements based on echocardiographic images as our inputs for above formula

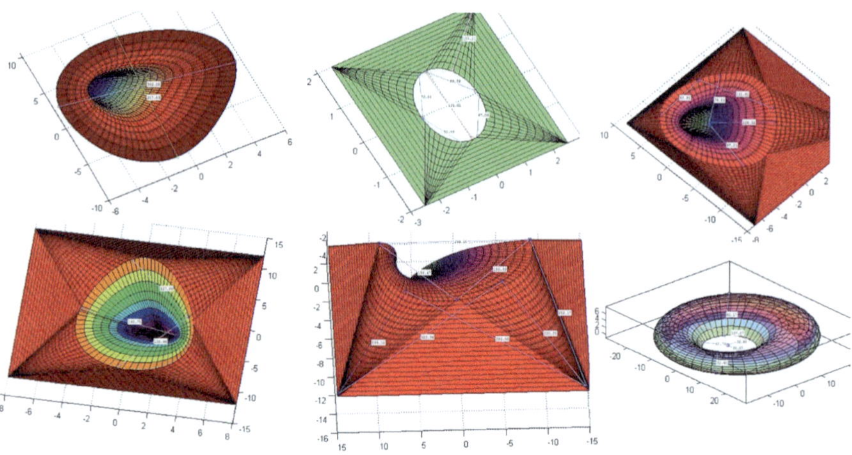

Fig. 13.3 3D reconstructions/measurements of the mitral valve based on Bernoulli's equation and the above formulae in the MATLAB software

Since $F\left(P,\dot{P},\varepsilon_{P,i},\dot{\varepsilon}_{P,i},t\right)=\partial\left(\varepsilon_{P,i}\times P_i\right)/\partial s_i$ (s_i the length of tangent line along P) we can also rewrite Lagrangian–Euler equations as:

$$\frac{\partial L\left(P_{i,}\dot{P}_i,\varepsilon_P,\dot{\varepsilon}_P,t\right)}{\partial P_i}-\frac{d}{dt}\left(\frac{\partial L\left(P_i,\dot{P}_i,\varepsilon_{P,i},\dot{\varepsilon}_{P,i,t}\right)}{\dot{P}_i}\right)=\partial\left(\varepsilon_{P,i}\times\dot{P}_i\right)/\partial s_i$$

$$\left[\left[\partial\left[\frac{1}{2}\rho\left(V_{\text{initial},p}+\det\begin{pmatrix}\varepsilon_x & \varepsilon_{xy} & \varepsilon_{xz}\\ \varepsilon_{yx} & \varepsilon_y & \varepsilon_{yz}\\ \varepsilon_{zx} & \varepsilon_{zy} & \varepsilon_z\end{pmatrix}\right)\dot{P}^2 -\frac{1}{2}\left(\sum_{\ell_{p,r}\in\ell^*}\left(\sum_{1\le i,j\le 3}e_{ij,\ell_{p,r}}(t)\cdot x_{i,r}\cdot x_{j,r}\right)\right)P^2\right]/\partial P_i\right]\right.$$

$$\left.-\left(d\left[\partial\left[\frac{1}{2}\rho\left(V_{\text{initial},p}+\det\begin{pmatrix}\varepsilon_x & \varepsilon_{xy} & \varepsilon_{xz}\\ \varepsilon_{yx} & \varepsilon_y & \varepsilon_{yz}\\ \varepsilon_{zx} & \varepsilon_{zy} & \varepsilon_z\end{pmatrix}\right)\dot{P}^2 -\frac{1}{2}\left(\sum_{\ell_{p,r}\in\ell^*}\left(\sum_{1\le i,j\le 3}e_{ij,\ell_{p,r}}(t)\cdot x_{i,r}\cdot x_{j,r}\right)\right)P^2\right]/\partial\dot{P}_i\right]/dt\right)=\partial\left(\varepsilon_{P,i}\times\dot{P}_i\right)/\partial s_i$$

Results

Conclusion

In conclusion, the results of this study suggest that Bernoulli's equation describe explicitly the behavior of the mitral valve leaflets with a huge clinical benefit for physicians. Our study also plays a significant role in the transvalvular pressure drop estimation of the mitral valve.

References

1. Brenke WC. An application of Abel's integral equation. Am Math Mon. 1922;29:58–60.
2. Bernoulli's equation is strictly valid for steady, frictionless flow in a stream tube. It is used, however, in engineering for flows with friction by modification of solutions with a suitable friction factor.

Discovery and the Development of the Mathematical Solution for Patient-Specific Human Heart Modelling Based on Echocardiographic Imaging

14

For decades automotive, aerospace, and energy industries have used advanced simulation technology to virtually design, test, and validate their products before they are constructed. Building on these successes, there has been a profound interest in medical sciences to use virtual simulation for designing, testing, and validating new treatment modalities. More specifically, this has focused on cardiovascular disease as it represents the primary cause of mortality in industrialized nations [1].

The heart is the most vital and complex organ of the human body and is controled by the interplay of anatomical, electrical and biomechanical events. In the past three decades, there has been a concerted effort in human heart modeling for facilitating clinical decision-making, guiding in treatment planning and acceleration in device design [2]. Despite these efforts, there remains a significant barrier in applying virtual models of the human heart in clinical applications. This is because the virtual simulation models are general engineering models based on a variety of medical images (anatomical data) combined with clinical measurements from a variety of invasive and non-invasive modalities that capture the biomechanical events for general multi-scale computational modeling [3, 4]. Most of these models have also been developed and validated using data from invasive measurements in controlled conditions in animal models [4]. These models are useful for demonstrating the proof of concept and development of general models. However, these models are clearly insufficient in terms of the ultimate objective of human heart modeling for the individualized prediction of different treatment modalities with the goal of virtually selecting the most promising treatment within the paradigm of personalized medicine.

In the case of patient-specific human heart modeling the challenge is to fuse the patient-specific geometrical data retrieved from an imaging modality with real-time in-plane biomechanical data of these geometrical points. In this chapter we report the discovery and the development of the mathematical solution for the retrieval and fusion of geometrical data with in-plane biomechanical data from any echocardiographic data set through the cardiac cycle of any anatomical point for patient-specific heart modeling.

© Springer Nature Switzerland AG 2023

M. Karvandi, S. Ranjbar, *A Review on Recent Echocardiographic Software*,

https://doi.org/10.1007/978-3-031-29046-6_14

The Concepts of the Mathematical Solution

Echocardiography emits high-frequency sound waves onto cardiac tissue, interpreting their reflection as images [5]. In fact, the scattering ultrasonic waves, the reversal waves, carry a significant amount of information with which to create images. These reversal waves provide data input for mathematical equations. After solving these equations, echocardiographic images are formed based on the Fourier series. In other words, an echocardiographic image is a solution of these equations of reversal waves. By putting these solutions together side by side, the movement of the heart is observed in the cardiac cycle and a video file is created.

Therefore, conceptually, an image or a frame in a cardiac phase corresponds to a set of solutions from the equations of reversal waves. For a simple understanding of the basis of the mathematical solution in this chapter, which is based on ultrasonic reversal waves, we will consider a set of solutions of wave equations as the images (a detailed mathematical equations is explained in the methodology). Hence, in the following general description of the mathematical solution, instead of a set of solutions of wave equations, we will use images as an analogy that corresponds to these solutions.

A raw two-dimensional echocardiography video is acquired. 2D raw images or frames are extracted. 2D echocardiographic video file consists of different frames and each frame consists of multiple pixels (Fig. 14.1). Each pixel represents an anatomical point in the heart that is tracked in a cardiac cycle. These anatomical points are represented geometrically by pixels on a two-dimensional Cartesian coordinate system (Fig. 14.1b). Because the points in the heart are moving, these pixels also move on the coordinate plane (Fig. 14.1a), and the study of the motion and deformation of these pixels corresponds to the movements of the anatomical points of the heart.

Based on the motion and deformation of anatomical points in the raw echocardiographic input data, the corresponding original velocity and strain of the pixels can be calculated on the Cartesian coordinate plane. According to Newton's second law, the calculation of the force that leads to motion and deformation is very

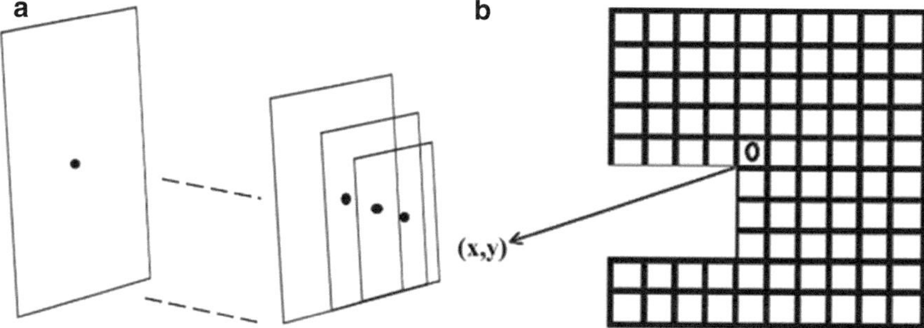

Fig. 14.1 (**a**) Movie is made by gluing the number of frames together. (**b**) Each pixel is considered in the Cartesian coordinate

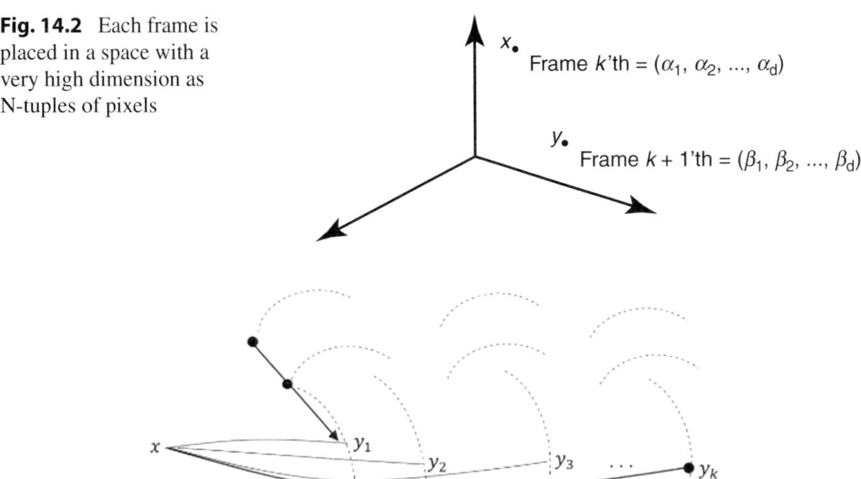

Fig. 14.2 Each frame is placed in a space with a very high dimension as N-tuples of pixels

Fig. 14.3 Suppose x represents a frame, with $y1$ being the closest point to x, the first neighbourhood being B1 (x) (a Sphere with center "x" and the radius with distance value between x, $y1$) and $y2$ being the closest point after $y1$ to point x, which makes the second neighbourhood B2 (x), and so on until neighbourhood k-th that we reach to the nearest anatomical point to x

important. Therefore, we formulate the force index in terms of velocity and strain (formulations with details were calculated, see methods). The initial amount of the force can be measured in terms of the initial velocity and strain obtained from velocity vector tracking. These values are entered into Lagrangian equations to obtain the trajectories achieved by the selected pixels according to Lagrange law in physics (mathematical details can be found in the methods).

Conversely, each frame consists of several pixels, for example N pixels. Therefore, each frame can easily be considered equivalent to an ordered N-tuples (Fig. 14.2). This shows that each frame can be placed in a Euclidean N-dimensional space.

The maximum and minimum values of the distances and the change of angles from the end of the diastole to the end of the systole are determined (tracking code). Therefore, we can draw spheres based on these values around the frames which are points in the N-dimensional space (Fig. 14.3).

Because the heart is elastic, the distance, angles, and deformation between the anatomical points in the heart do not change in a cardiac cycle. We can define a mapping "f" (f_p's) from this N-dimensional space and the spheres drawn into the 3D Cartesian coordinate space so that the values of this mapping are compatible with the trajectories obtained by solving the Lagrangian equations. This mapping **f** (Fig. 14.4), which is based on pixel tracking and the trajectories obtained by the Lagrange equations, provides precise absolute in-plane real-time biomechanical parameters of any anatomic point in the heart during the full cardiac cycle, such as velocities, strains, forces, divergences, vortices, lengths, distances, angels and a fiber bundle of l_p 's that represents flow movements (optimized trajectories) of each

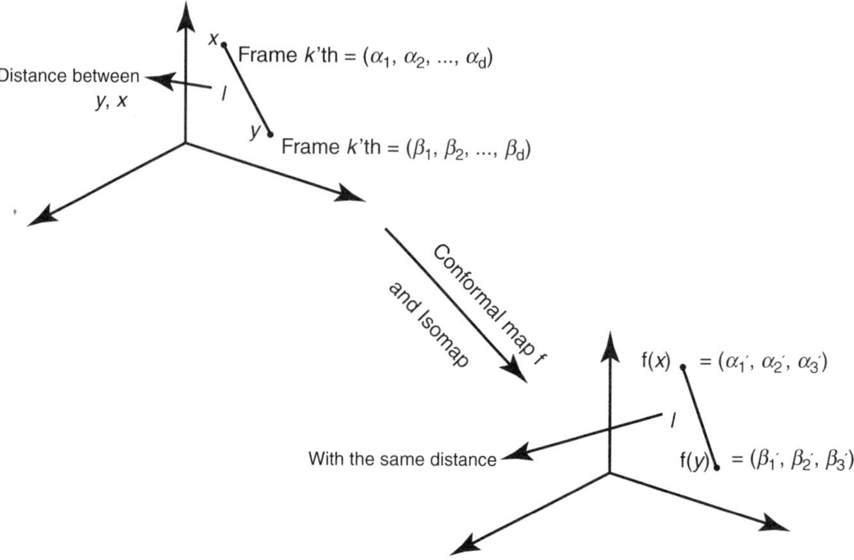

Fig. 14.4 The mapping "**f**" from N-dimensional space to a 3D Cartesian space that preserve distances (Iso-map f) and angels (conformal map f) and compatible with trajectories obtained by the Lagrange equations

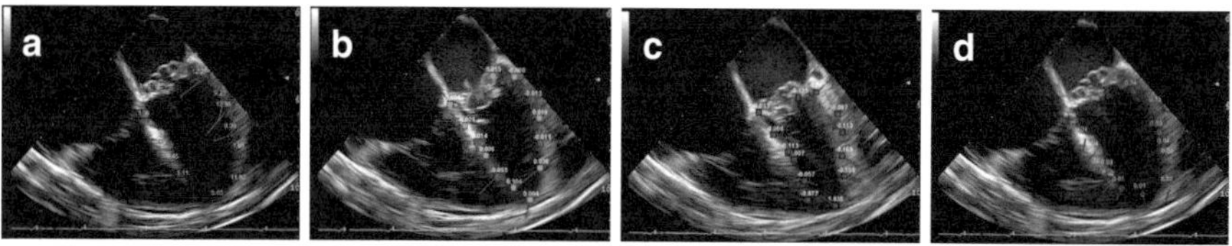

Fig. 14.5 (**a–d**) (**a**) Vector velocities with numerical values (mm/s) are fused with geometrical points through cardiac cycle. (**b, c**) Strain vectors and their numerical values coded with colour (red states negative values and green color represents the positive values). (**d**) Force vectors and their numerical values (millimeter Newton).

cardiac points 'p' in the heart per a single cardiac cycle. Furthermore, these precise in-plane mechanical events are fused with geometrical points (Fig. 14.5).

For the 3D mode (Fig. 14.6), a voxel replaces instead of a pixel and the mathematical workflow is the same as above, with the key difference that mapping occurs in a Cartesian 3D space which preserves rotational indices and compatibles with Lagrange equations in terms of 3D motion and strain data (mathematical details can be found in the methods). The additional consequence of this model is that one can retrieve an unlimited number of frames and volumes between existing frames and volumes visualized in echocardiographic images. Furthermore, using K-theory in the algebraic geometry, any structure of the heart can be tracked as a separate entity for virtual simulation.

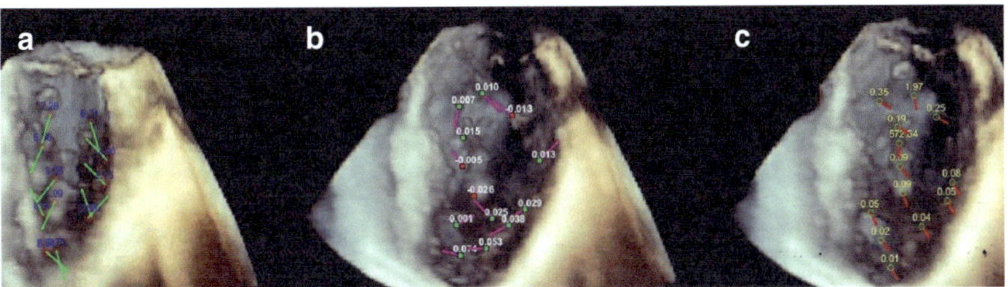

Fig. 14.6 (**a**) 13 points are selected in left in 3D left ventricle full volume and velocity vectors with green colour and their numerical values attached to the selected points with mm/s unit. (**b**) 13 points are selected and strain vectors with red (negative value) and green (positive value) colour coded. (**c**) 13 points are selected and force vectors with red colour that are resulted in those velocity and strain with their numerical values attached to the selected points with mmN (millimeter Newton) unit

Programming and Testing

All mathematical algorithms were coded with more than 50,000 computer commands. The C++ computer language was used; thus, codes are convertible to lower-level languages to get better performance and could be integrated into the libraries in enterprise software packages. These algorithms work independent of any imaging platform, meaning that one can retrieve all these data from any existing echocardiographic dataset.

We tested our program with datasets from different echocardiography machines, including Philips (EPIQ CVx and iE33 xMATRIX), GE Healthcare (Vivid 3, Vivid 7 and LOGIQ E9) and Esaote (MyLab™ 60, MyLab™ 70).

In another case we applied the mathematical algorithms on a 3D surgical view of the mitral valve. The original frames of 12 were increased to 120. We traced 60 points on the annulus of the mitral valve and tracked the velocity vectors throughout the cardiac cycle. By using K-theory in the algebraic geometry different structures of the mitral valve could be tracked throughout the cardiac cycle (Fig. 14.7).

In another case we applied the mathematical algorithms on a long-axis view of the heart on 2D TEE. Here we increased the number of frames from 66 to 660. Over 120 points were selected for which the velocity vectors were tracked throughout the cardiac cycle. By using K-theory methods in the algebraic geometry, we could track the mitral apparatus including the native chords throughout the cardiac cycle (Fig. 14.8).

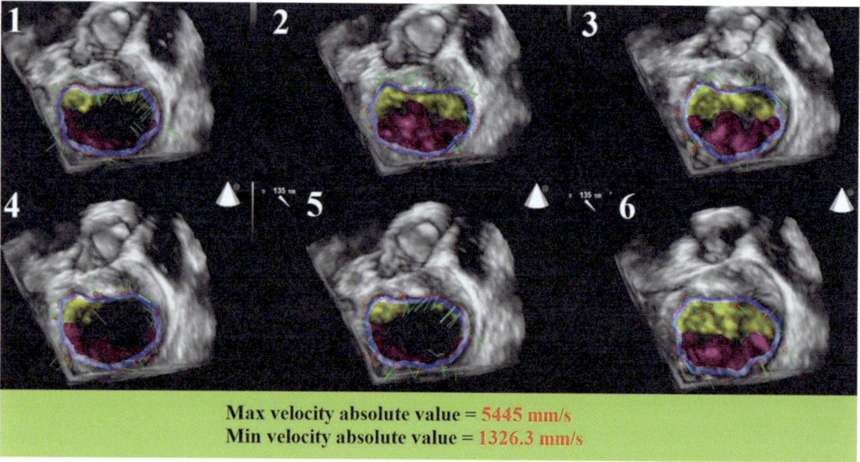

Fig. 14.7 Mathematical imaging of mitral valve based on 3D TEE: the original number of volumes was 12, which we increased to 120. (*1–6*) sixty points (red) were traced on the annulus of the mitral valve. And, similarly, we selected 60 points on the mitral valve leaflets. The annulus and mitral valve leaflets were tracked by velocity vectors during the cardiac cycle. Different velocity vectors directions; shortening and lengthening are visible of each point in the annulus of the mitral valve. The absolute values of velocity were between 5445 mm/s and 1326.3 mm/s. Utilizing K-theory in algebraic geometry and the velocity vector fields of the mitral valve annulus and the mitral leaflets, we could distinguish with color the mitral valve annulus (blue), the posterior mitral valve leaflet (pink) and the anterior mitral valve leaflet (yellow) during the full cardiac cycle

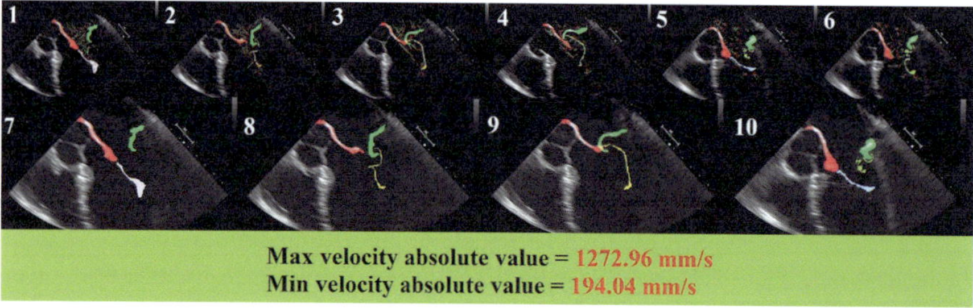

Fig. 14.8 Mathematical imaging of mitral valve apparatus based on 2D TEE: The number of 66 original frames was increased to 660 frames. Many pixels and phases were extracted within a cardiac cycle. (*1–7*) 120 points (red) were traced on the mitral valve leaflets and behind the P2 prolapse in the left ventricular side. Different velocity vectors directions; shortening and lengthening are visible point by point. The absolute values of velocity were between 1272.96 mm/s and 194.04 mm/s. (*8–10*) Using K-theory methods in the algebraic geometry, we were able to track anterior mitral valve leaflet (pink color), posterior mitral valve leaflet (green color), elongated chordae of a chordae of posterior leaflet (yellow) and chordae of the anterior mitral valve (white)

Discussion

The vast majority of human heart models are general engineering models based on a variety of medical images (anatomical data) combined with clinical measurements from a variety of invasive and non-invasive modalities that capture the biomechanical events for multi-scale computational modelling [2, 3]. Most of these models have also been developed and validated using data from invasive measurements in controlled conditions and do not provide a patient-specific solution [4]. In the case of patient-specific human heart modeling, the challenge has been to retrieve and fuse the patient-specific geometrical points with absolute real-time in-plane mechanical data from patients' images.

Here, we have discovered and developed the mathematical solution for the retrieval and fusion of geometrical data with in-plane biomechanical data from any echocardiographic data set through a cardiac cycle for patient-specific heart modeling. Our mathematical algorithms also increase the quality and quantity of the echocardiographic source data by locating more anatomical points (pixels or voxels) in time and space, thereby improving the temporal and spatial resolution of original datasets. This solution allows us to retrieve information about the dynamics of selected anatomical points through a cardiac cycle, namely the precise absolute in-plane real-time mechanical and dynamical parameters such as velocities, strains, forces, divergences, vortices, lengths, distances, angels and a fiber bundle of l_P' s that represents flow movements (optimized trajectories) of each cardiac points 'p' in the heart per cardiac cycle. Our algorithm is independent of any imaging platform and can process data originating from all existing imaging machines. Furthermore, using K-theory in the algebraic geometry any structure and region of the heart can be tracked as a separate entity for virtual simulation.

Ultrasonic frequencies pulsed into cardiac segments by advanced probes range from a low of 10 MHz (rarely, 5 MHz) to a high of over 400 MHz. In this range of frequencies there is a trade-off between penetration and resolution. Low-frequency ultrasound infiltrates materials deeper than higher frequencies, but the spatial resolution of the image is low. Conversely, very high-frequency ultrasound does not penetrate as deeply, but it does, however, provide acoustic images with higher resolution [5]. The frequency selected to visualise a particular anatomical region also depends on the geometry, motion, and deformation of the scanned tissue [6]. Overcoming this limitation is one of the key applications of our algorithm.

Tracking any structure in the heart and obtaining the geometrical changes of these structures is possible in our algorithm. The velocities, strains, forces, divergences, vortices, lengths, distances, angels, and a fiber bundle of l_P vectors from different anatomical points provide insight into physiological changes during the cardiac cycle that could help our understanding of the physiology and pathology of different cardiac structures.

The main advantage of our approach is that we can reconstruct an environment based on advanced mathematical techniques effective on any reversal equation. An additional advantage of applying mathematics directly to the source data is that the concepts applied here can be used on any imaging modality, with potential use in calculating many biomechanical parameters, image fusion, artificial intelligence and virtual simulation [7–9].

In artificial intelligence, image classification is used for machine learning. Large collections of images are used to train, and then evaluate, a classifier using new images [10–12]. By applying mathematical algorithms to diagnostic images, we may be able to formulate a mathematical template for a given disease or condition, which would greatly enhance the application of artificial intelligence in medicine.

Our proposed method can also be used in myocardial strain calculations. The strain parameter is measured using different mathematical techniques that are indirect measurements and extrapolations with operator dependency such as: tissue Doppler imaging combined with velocity gradient, speckle tracking, speckle tracking-end 7 points, speckle tracking combined with tissue Doppler imaging, and 2D base strain (speckle tracking combined with velocity vector imaging); ventricular chamber modeling as a polygon shape combined with speckle tracking. Even on a simple M-mode one is able to measure some kind of strain. Therefore, the value depends on the kind of strain (longitudinal, radial, circumferential, etc.) and how it is measured. To achieve a high level of accuracy, our algorithm can be used on any echocardiographic platform for measuring velocity and deformation indices [13]. The images obtained by different systems could be input into our algorithms to reconstruct new images with greater detail (revealing additional hidden data), and subsequently used to measure multiple parameters by implementing the mathematical techniques described.

The development of new algorithms for other imaging modalities like MRI opens a whole new field where we are able to create new imaging modalities to replace existing invasive techniques.

Mathematical Methods

Heart (H) is considered mathematically as a 3D + t = 4D manifold based on the Lagrangian mechanical system [14], as a biological circulatory system corresponds to a 4D manifold in the space and time.

Anatomical points ('p') correspond to the geometrical points of a 3D + t = 4D smooth manifold H. Each geometrical point moves in a closed time interval during the cardiac cycle. Velocity vector fields are considered phase by phase, resulting in a bundle of velocity vectors $v_{p,t}$ (Fig. 14.9).

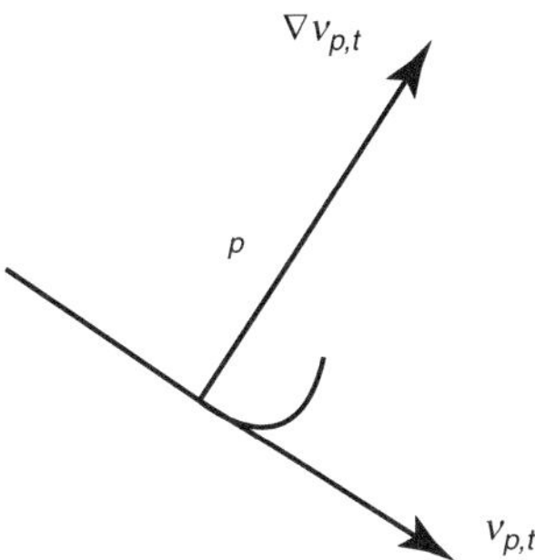

Fig. 14.9 A cardiac point p at the phase t with the velocity vector $v_{p,t}$ and gradient vector $\nabla v_{p,t}$.

V_t is the space of velocity vectors attached at time t. Based on the F-sheaves theory [15], we have the space of 3 by 3 matrices through the time in a cardiac cycle as follows:

$$H_p(t) := \left(v_{p,t}\Delta\cdot\nabla v_{p,t}\right)v_{p,t} = \begin{matrix} v_{p,t} & v_{p,t,x} & v_{p,t,y} & v_{p,t,z} \\[4pt] & \dfrac{\partial v_{p,t,x}}{\partial x}v_{p,t,x} & \dfrac{\partial v_{p,t,y}}{\partial y}v_{p,t,y} & \dfrac{\partial v_{p,t,z}}{\partial z}v_{p,t,z} \\[8pt] \left(v_{p,t}\times\nabla v_{p,t}\right)v_{p,t} & \dfrac{\partial v_{p,t,y}}{\partial y}-\dfrac{\partial v_{p,t,x}}{\partial x} & \dfrac{\partial v_{p,t,z}}{\partial z}-\dfrac{\partial v_{p,t,x}}{\partial x} & \dfrac{\partial v_{p,t,y}}{\partial y}-\dfrac{\partial v_{p,t,z}}{\partial z} \end{matrix}$$

The first row describes the velocity components, the second row shows the divergence at the point p and the third row states the vortices at the point p phase by phase within a cardiac cycle. Therefore, H is realized as a stack of smooth 3D manifolds (cross-sections of H) through the cardiac cycle toward the velocity vector fields.

The 3D strains are presented as 3 by 3 matrices in the following manner:

$$\varepsilon_{p,t} = \begin{bmatrix} \varepsilon_{p,t,11} & \cdots & \varepsilon_{p,t,13} \\ \vdots & \ddots & \vdots \\ \varepsilon_{p,t,31} & \cdots & \varepsilon_{p,t,33} \end{bmatrix}$$

Force indices are formulated and calculated based on extracted motion and deformation parameters by the following tensor product:

$$F_{p,t} = H_p(t)\otimes\varepsilon_{p,t} = \begin{matrix} v_{p,t,x} & v_{p,t,y} & v_{p,t,z} \\[4pt] \dfrac{\partial v_{p,t,x}}{\partial x}v_{p,t,x} & \dfrac{\partial v_{p,t,y}}{\partial y}v_{p,t,y} & \dfrac{\partial v_{p,t,z}}{\partial z}v_{p,t,z} \\[8pt] \dfrac{\partial v_{p,t,y}}{\partial y}-\dfrac{\partial v_{p,t,x}}{\partial x} & \dfrac{\partial v_{p,t,z}}{\partial z}-\dfrac{\partial v_{p,t,x}}{\partial x} & \dfrac{\partial v_{p,t,y}}{\partial y}-\dfrac{\partial v_{p,t,z}}{\partial z} \end{matrix}$$

$$\otimes \begin{bmatrix} \varepsilon_{p,t,11} & \cdots & \varepsilon_{p,t,13} \\ \vdots & \ddots & \vdots \\ \varepsilon_{p,t,31} & \cdots & \varepsilon_{p,t,33} \end{bmatrix}$$

General force of p per cardiac cycle is the summation of the above formula:

$$F_P = \underset{t \text{ runs in a cardiac cycle}}{\oplus} F_{p,t} = \underset{t \text{ runs in a cardiac cycle}}{\oplus}$$

$$v_{p,t,x} \qquad v_{p,t,y} \qquad v_{p,t,z}$$

$$\frac{\partial v_{p,t,x}}{\partial x} v_{p,t,x} \qquad \frac{\partial v_{p,t,y}}{\partial y} v_{p,t,y} \qquad \frac{\partial v_{p,t,z}}{\partial z} v_{p,t,z}$$

$$\frac{\partial v_{p,t,y}}{\partial y} - \frac{\partial v_{p,t,x}}{\partial x} \qquad \frac{\partial v_{p,t,z}}{\partial z} - \frac{\partial v_{p,t,x}}{\partial x} \qquad \frac{\partial v_{p,t,y}}{\partial y} - \frac{\partial v_{p,t,z}}{\partial z}$$

$$\otimes \begin{bmatrix} \varepsilon_{p,t,11} & \cdots & \varepsilon_{p,t,13} \\ \vdots & \ddots & \vdots \\ \varepsilon_{p,t,31} & \cdots & \varepsilon_{p,t,33} \end{bmatrix}$$

The Lagrangian equation for 3D cardiac segment (point p) is described by the following reformulation based on the acquired data sets:

$$L\left(p,\dot{p},\varepsilon_P,\dot{\varepsilon}_P,t\right) = T\left(p,\dot{p},\varepsilon_{P,t},\dot{\varepsilon}_{P,t}\right) - U\left(p,p,\varepsilon_{P,t},\dot{\varepsilon}_P,t\right)$$

$$= \frac{1}{2}\rho\left[\delta Volume_{p,n}\left(t_n\right) + p_1 p_2 p_3\right]\dot{p}^2 - \frac{1}{2}\left(\sum_{\ell_{p,r} \in \ell_p^*}\left(\sum_{1 \le i,j \le 3} \varepsilon_{p,t,ij,\ell_p} \cdot P_i \cdot P_j\right)\right)p^2$$

where: ℓ_p^* is the set of all fibers passing through p; $\delta Volume_p(t)$ is the change in volume of the point p from the first volume to n's volume in time t_n; and $\delta Volume_{p,n}(t_n)$ is the determinant of strain matrix:

$$\varepsilon_{p,t} = \begin{bmatrix} \varepsilon_{p,t,11} & \cdots & \varepsilon_{p,t,13} \\ \vdots & \ddots & \vdots \\ \varepsilon_{p,t,31} & \cdots & \varepsilon_{p,t,33} \end{bmatrix}$$

Based on this, Lagrangian equations are reformulated:

$$\frac{\partial L\left(p,\dot{p},\varepsilon_{P,t},\dot{\varepsilon}_P,t\right)}{\partial p} - \frac{d}{dt}\left(\frac{\partial L\left(p,\dot{p},\varepsilon_{p,t},\dot{\varepsilon}_p,t\right)}{\partial \dot{p}}\right) = F_P = \underset{t \text{ runs in a cardiac cycle}}{\oplus}$$

$$F_{p,t} = \underset{t \text{ runs in a cardiac cycle}}{\oplus} \qquad v_{p,t,x} \qquad v_{p,t,y} \qquad v_{p,t,z}$$

$$\frac{\partial v_{p,t,x}}{\partial x} v_{p,t,x} \qquad \frac{\partial v_{p,t,y}}{\partial y} v_{p,t,y} \qquad \frac{\partial v_{p,t,z}}{\partial z} v_{p,t,z}$$

$$\frac{\partial v_{p,t,y}}{\partial y} - \frac{\partial v_{p,t,x}}{\partial x} \qquad \frac{\partial v_{p,t,z}}{\partial z} - \frac{\partial v_{p,t,x}}{\partial x} \qquad \frac{\partial v_{p,t,y}}{\partial y} - \frac{\partial v_{p,t,z}}{\partial z}$$

$$\otimes \begin{bmatrix} \varepsilon_{p,t,11} & \cdots & \varepsilon_{p,t,13} \\ \vdots & \ddots & \vdots \\ \varepsilon_{p,t,31} & \cdots & \varepsilon_{p,t,33} \end{bmatrix}$$

By solving these equations, we provide curves l_P's point by point. In fact, a fiber bundle of curves l_P's is constructed. This fiber bundle represents flow movements (optimized trajectories) of each cardiac point per a cardiac cycle.

Addition of Gaussian Curvatures, Connections, Riemannian Curvatures, and Hamilton-Ricci [16–18] Flow Equation

For a cardiac point 'p' we obtained the optimized trajectory l_P. **We set the Gaussian curvature point 'p' along the** optimized trajectory l_P **by the following formula:**

$$r_{(l_P)} = \text{The Gussian curvature at the point } 'p'$$

$$\tan \alpha_{l_P} = \left\| l_P \right\| \rightarrow\rightarrow \alpha_{l_P} = \arctan\left(\left\| l_P \right\| \right)$$

$$r_{l_P} = \int_{p \text{ runs away on the cardiac points}} \qquad \alpha_{l_P} = \int_{p \text{ runs away on the cardiac points}} \arctan\left(\left\| l_P \right\| \right)$$

H as a 4D smooth manifold is divided to H_1 and $H_2 \cdot H_1$ is constructed based on Lagrangian equations and H_2 is constructed by utilizing Hamilton-Ricci flow equations.

Whenever a cardiac point 'p' moves along the optimized trajectory l_P, for instance, from phase t_1 with the velocity vector $v(p,t_1)$ and stain ε_{p,t_1} to the next phase t_2 with the velocity vector '$v(p,t_2)$'. Vectors '$v(p,t_1)$' and '$v(p,t_2)$' connect by something like a stent (Fig. 14.10). The connection is defined by the following manner:

$$\nabla_{v(p,t_1)} = \varepsilon_{p,t_1} \times v_{p,t_1} = \begin{bmatrix} \varepsilon_{p,,t_1 11} & \cdots & \varepsilon_{p,,t_1 13} \\ \vdots & \ddots & \vdots \\ \varepsilon_{p,,t_1 31} & \cdots & \varepsilon_{p,t_1 ,33} \end{bmatrix} = v\left(p,t_2\right)$$

On the other hand, vector '$v(p,t_1)$' is connected to the vector '$v(p,t_2)$' by the stent ε_{p,t_1} after passing time $t_2 - t_1$.

Fig. 14.10 For a cardiac segment 'p' during two different phases, vector '$v(p,t_1)$' is connected to the vector '$v(p,t_2)$' by the stent ε_{p,t_1} after passing time $t_2 - t_1$.

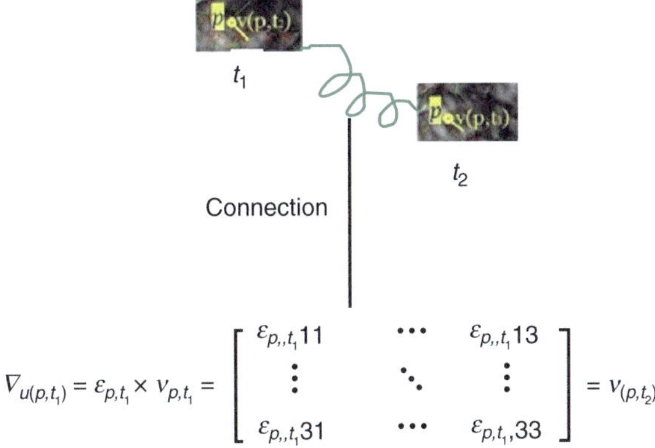

$$\nabla_{u(p,t_1)} = \varepsilon_{p,t_1} \times v_{p,t_1} = \begin{bmatrix} \varepsilon_{p,,t_1}11 & \cdots & \varepsilon_{p,,t_1}13 \\ \vdots & \ddots & \vdots \\ \varepsilon_{p,,t_1}31 & \cdots & \varepsilon_{p,t_1,}33 \end{bmatrix} = v_{(p,t_2)}$$

We have these connections phase by phase per a cardiac cycle. Therefore, the Riemannian curvature 'g' is reformulated in the following way:

$$g(u,v)(w) = \nabla_u \nabla_v w - \nabla_v \nabla_u w$$

The Hamilton-Ricci flow eqs. $\dfrac{dg}{dt} = (r - R)g$ state the bending behavior of $_{\text{Heart}}$ (H), where r is the whole Gaussian curvature and R is the average of r_{l_p}'s. Finally, based on the echocardiography, a set of partial differential equations (PDEs) were designed based on all obtained equations, to construct a mathematical heart model H [19]. The set of solutions H of these PDEs were calculated numerically using spectral methods. The set of all transformations on H was extracted based on Lie theory [20]. By this novel modeling we can extract from patient-specific images, real-time in-plane mechanical parameters from any anatomic points that will form the basis for patient-specific heart modeling.

An Explanation of the Human Heart Mathematical Modeling

Utilizing the same arguments in above and K-theory [21] with heart composed in different structures as a unitary reductive group [22], one can reconstruct the different structures of the heart in a lot of cardiac phases. One can extract all mechanical and dynamical parameters from these structures in a cardiac cycle for the solution of Lagrangian equations. One can also reformulates the connections point by point in different phases and Riemannian curvature to solve the Hamilton-Ricci flow equations. This method and material give a mathematical modelling of different structures of the heart (Figs. 14.7 and 14.8).

References

1. Virani SS, et al. Heart disease and stroke statistics-2020 update: a report from the American Heart Association. Circulation. 2011;141(9):e139–596.
2. Peirlinck M, et al. Precision medicine in human heart modeling: perspectives, challenges, and opportunities. Biomech Model Mechanobiol. 2021;20(3):803–31.
3. Baillargeon B, et al. The living heart project: a robust and integrative simulator for human heart function. Eur J Mech A Solids. 2014;48:38–47.
4. Smith N, et al. euHeart: personalized and integrated cardiac care using patient-specific cardiovascular modelling. Interface Focus. 2011;1(3):349–64.
5. Armstrong WF, Ryan T. Feigenbaum's echocardiography. 8th ed. Baltimore: Lippincott Williams & Wilkins; 2018.
6. Errico C, et al. Ultrafast ultrasound localization microscopy for deep super-resolution vascular imaging. Nature. 2015;527(7579):499–502.
7. Alberti GS, et al. Mathematical analysis of ultrafast ultrasound imaging. SIAM J Appl Math. 2017;77:1–25.
8. Madani A, et al. Fast and accurate view classification of echocardiograms using deep learning. NPJ Digit Med. 2018;1:6.
9. Christensen-Jeffries K, et al. 3-D *in vitro* acoustic super-resolution and super-resolved velocity mapping using microbubbles. IEEE Trans Ultrason Ferroelectr Freq Control. 2017;64:1478–86.

10. Marchesseau S, et al. Cardiac mechanical parameter calibration based on the unscented transform. In: Ayache N, et al., editors. Medical image computing and computer-assisted intervenion-MICCAI. Lecture notes in computer science, vol. 7511. Cham: Springer; 2012. p. 41–8.
11. Piccinelli M, Garcia E. Multimodality image fusion for diagnosing coronary artery disease. J Biomed Res. 2013;27:439–51.
12. Ouyang D, et al. Video-based AI for beat-to-beat assessment of cardiac function. Nature. 2020;580(7802):252–6. https://doi.org/10.1038/s41586-020-2145-8.
13. Narula S, et al. Machine-learning algorithms to automate morphological and functional assessments in 2D echocardiography. J Am Coll Cardiol. 2016;68:2287–95.
14. Arnold VI. Mathematical methods of classical mechanics. New York: Springer; 1997.
15. Drinfel'd VG. Varieties of modules of F-sheaves. In: Functional analysis and its applications, vol. 21; 1987. p. 107–22.
16. Lee JM. Riemannian manifolds: an introduction to curvature. Cham: Spinger; 1997.
17. Chow B, Knopf D. The ricci flow: an introduction. Mathematical surveys and monographs, vol. 110. Providence: American Mathematical Society; 2004.
18. Chow B, et al. The ricci flow: techniques and applications. Part I. geometric aspects. Mathematical surveys and monographs, vol. 135. Providence: American Mathematical Society; 2007.
19. Schoen R, Yau S-T. Proof of the positive mass theorem. II. Commun Math Phys. 1981;79:231–60.
20. Olver PJ. Applications of lie groups to differential equations. Cham: Springer; 1986.
21. Atiyah MF, Anderson DW. K-theory. 1st ed. Boca Raton: CRC; 1967.
22. Nevine M, Trapa PE. Representation of reductive lie group. Cham: Springer; 2015.

Index

© Springer Nature Switzerland AG 2023
M. Karvandi, S. Ranjbar, *A Review on Recent Echocardiographic Software*,
https://doi.org/10.1007/978-3-031-29046-6

If you have any concerns about our products,
you can contact us on
ProductSafety@springernature.com

In case Publisher is established outside the EU,
the EU authorized representative is:
Springer Nature Customer Service Center GmbH
Europaplatz 3, 69115 Heidelberg, Germany

Printed by Libri Plureos GmbH
in Hamburg, Germany